# Advanced Classical Mechanics

# Advanced Classical Mechanics

Bijan Kumar Bagchi

CRC Press
Taylor & Francis Group
Boca Raton London New York

CRC Press is an imprint of the
Taylor & Francis Group, an **informa** business

CRC Press
Taylor & Francis Group
6000 Broken Sound Parkway NW, Suite 300
Boca Raton, FL 33487-2742



Printed on acid-free paper
Version Date: 20170330

International Standard Book Number-13: 978-1-4987-4811-7 (Hardback)



**Visit the Taylor & Francis Web site at**
**http://www.taylorandfrancis.com**

**and the CRC Press Web site at**
**http://www.crcpress.com**

*In my father's memory*

# Contents

# *Preface*

The book has evolved from my teaching of the subject of classical mechanics for many years. It is designed to serve as a textbook for the young postgraduate students, researchers and teachers of theoretical physics/applied mathematics and also students of engineering who need a good background and understanding of classical mechanics. We have therefore placed emphasis on the logical ordering of topics and appropriate formulation of the key mathematical concepts with a view to imparting a clear knowledge of the basic tools in the subject and improving the problem solving skills of the students. Complexities of the mathematical structure are kept to a minimum but greater stress is laid on the application-side of the subject. We have tried to develop new ideas as a smooth continuation of the preceding ones. The book is expected to give a systematic and comprehensive coverage of the methods of classical mechanics.

The book is organized into nine chapters and begins, in Chapter 1, with the conceptual basis of classical mechanics to provide the necessary background for the later chapters. This is expected to enable the students to have a grasp on the perspectives behind the development of classical mechanics. In Chapter 2 a treatment of central force problems is presented and certain typical issues like Kepler's laws, power law potentials and Laplace–Runge–Lenz vector are studied in much detail. Chapter 3 is concerned with the treatment of Lagrangian dynamics. A large variety of problems is solved in this chapter. Considerable emphasis is placed on the basic ideas keeping in mind the difficulties that a student could face when being exposed to these principles for the first time. Brief discussions of ignorable coordinates and Liouville's class of problems are given here. We have also included the topic of small oscillations. In Chapter 4 a formulation of Hamiltonian dynamics is developed and Hamilton's equations of motion are derived. Also discussed in this chapter are the Poisson brackets, their various properties and utilities. The case of singular Lagrangians is dealt with in detail and a section is devoted toward the treatment of higher derivative Lagrangians. Chapter 5 considers an overview of dynamical systems and covers topics such as an analysis of linear systems, Lotka-Volterra models, Lyapunov systems, van der Pol oscillator, limit cycles and the theory of bifurcations. Chapter 6 focuses on a detailed treatment of action principles and discusses extended point transformations along with different types of variations. The brachistochrone problem is studied here. Rotating frames and velocity dependent potentials are introduced in Chapter 7 along with a treatment of the non-potential force. Its relevance in physical phenomena is highlighted. The topic of Foucault's pendulum is also touched upon. Chapter 8 takes up the role of symmetries and conservation laws in mechanical systems. A discussion of the operator approach is given.

Chapter 9 is concerned with the canonical transformations and their role in physical problems. Topics like the Hamilton–Jacobi equation and the theory of action-angle variables are treated extensively. A notable feature of this book is that each chapter contains a wide range of worked out problems to clarify the basic ideas involved. It is believed these will help further the understanding of the subject.

First I thank Prof. Rupamanjari Ghosh, Vice Chancellor of Shiv Nadar University for sustained encouragement during the entire process of writing this book. I am also obliged to thank Prof. Sankar Dhar, head of the Department of Physics, School of Natural Sciences, Shiv Nadar University for excellent cooperation toward successful completion of the book.

It was the late Jnanendra Gopal Chakraborti who introduced me to the subject of classical mechanics and greatly impressed me with his unique teaching methods. I am indebted to him.

I profoundly thank Dr. Anindya Ghose Choudhury and Dr. Priya Johari for reading certain portions of the manuscript and giving their insightful remarks and candid opinions. I express my sincere appreciation to Dr. Santosh Singh for his help in finalizing the draft and kind interest. I am thankful to my colleagues Prof. Tanuka Chattopadhyay, Dr. Soumen De, Prof. Samiran Ghosh, Prof. Partha Guha, Dr. Swarup Poria, and Prof. Arabinda Roy for their comments at various stages. Thanks are also due to many of my students for their assistance during the writing of the book. I take this opportunity to make a special mention of Dr. Abhijit Banerjee, Mr. Partha Mandal, Ms. Saheli Mitra, and Mr. Tarun Tummuru. The references listed at the end have deeply influenced me in making up the material of the book and I recommend them for further reading.

I gratefully acknowledge the assistance from the librarians of the Inter-University Center for Astronomy and Astrophysics, Pune and the Indian Institute of Technology, Kanpur for access to their libraries and providing a convenient ambiance for work.

I would also like to express my appreciation to Ms. Aastha Sharma, Commissioning Editor, CRC Press, Taylor & Francis Group for not only helping me to initiate the project but also for her excellent cooperation in turning the book into a reality.

I owe thanks to the late Jayanta Kumar Bagchi, my uncle, who constantly urged me to complete the book but unfortunately did not live to see its publication.

Finally, I thank my wife Minakshi, and daughter Basabi, for their patience and support.

Shiv Nadar University
Greater Noida

Bijan Kumar Bagchi

# Chapter 1

# Conceptual basis of classical mechanics



Classical mechanics is the study of physical laws that control the motion of material objects which are under the action of a force or system of forces. It provides the basis for the growth of modern science. It has applications that cover areas such as physics, chemistry, applied mathematics, biology and engineering sciences. In particular, it seeks to address and explain the dynamics of particles and rigid bodies, general classes of interactive systems, rotating Earth problems, motion of charged objects, planetary motions around the Sun and modeling of biological systems. Classical mechanics has an extraordinarily rich history that began about the time of Galileo (1564–1642) although its basic foundations were laid later by Newton (1642–1727) in his famous treatise, the *Principia*. He enumerated a set of three axioms which became the cornerstone in explaining most of the qualitative features of classical mechanics. In this chapter we discuss these laws and provide a multi-layered perspective on them.

## 1.1 Newton's three laws

Newton's three laws may be stated as follows:

1. Every body continues in its state of rest or of uniform motion in a straight line unless it is compelled by external forces to change the state.

2. The rate of change of momentum is proportional to the impressed force and takes place along the direction in which the force acts.

3. To every action there is an equal and opposite reaction.

The first law provides the concept of inertia. It is also referred to as the Law of Inertia. The idea of absoluteness of space (Euclidean) and time is inherent in the Newtonian scheme of things. Both are looked upon as independent entities. As Newton noted in his *Principia*: "*Absolute space, in its own nature, without regard to anything external, remains always similar and immovable.*" About absoluteness of time, which is measured vis-a-vis the motion of different objects and which evolves continuously, Newton opined that "*Absolute, true, and mathematical time, of itself, and from its own nature, flows equably without regard to anything external, and by another name is called duration.*" Despite the independence of space and time, their correlation cannot be denied in that for a material point characterized only by its location without any reference to time is not of much significance. Conversely, the happening of an event at a particular point of time or for a certain duration with *a priori* no knowledge of its location looks conceptually absurd.

There are actually two aspects to the first law that may be necessary to be kept in mind. It at once seeks to address both a stationary object and a moving one. For the stationary object it says that an object at rest continues to do so unless acted upon by an external force. For a moving object the law holds that a particle in motion remains in motion (in a straight line) with the same speed and direction unless of course it is interacted upon by an external force.

Newton's first law also defines a class of reference frames which are called inertial frames. These reference frames are not supposed to undergo any type of acceleration. In other words, an inertial frame moves with a constant velocity or is at rest with respect to any other inertial frame. In 1886 Ludwig Lange, a German physicist, defined an inertial frame to be: "*A reference frame in which a mass point thrown from the same point in three different non-coplanar directions follows rectilinear paths each time it is thrown, is called an inertial frame.*" An inertial frame may be looked upon as the one that is fixed relative to the average position of a fixed star or that is moving with a constant velocity and without any rotation relative to it. According to Newton, absolute space and time form a convenient background against which we visualize occurrence of physical phenomena but the background itself cannot be influenced by physical events themselves. To describe a physical event, a fixed reference coordinate frame (inertial) must be chosen with respect to which the location of a material object is defined.

A non-inertial frame is the one that is not inertial. A rotating frame is a common example of a non-inertial frame. Although for short scales of time and distance we can consider the planet Earth to be an inertial frame, truly speaking, it is rotating about an axis passing through its geographical poles. Hence, technically, it does not constitute an inertial frame. A rotating frame, as we will find later, generates fictitious or pseudo forces. The Coriolis force is such a type of force and we need to carefully deal with it along with the so-called centrifugal force to make a proper use of Newton's laws in rotating frames.

A physical event with coordinates $\vec{r} = (t, \vec{x})$ is confined to the direct product of the $t$-axis with the time parameter $t \in \mathbb{R}$ and space variables $\vec{x} \in \mathbb{R}^3$. The latter is equipped with a fixed Euclidean structure called the Galilean space. Such a set of space-time coordinates signifies an event.

Consider specifically a one-dimensional motion in which the $x$-coordinates of two observers define two different moving frames $S$ and $S'$. Let $|\vec{u}|$ denote the magnitude of uniform velocity of $S'$ relative to $S$ along their common $x$ and $x'$ axis with the same spatial origin coinciding at time $t = t' = 0$. Due to the inherent inertial character Newton's laws are unaffected by transformations of the type

$$x' = x + ut, \quad y' = y, \quad z' = z, \quad t' = t \tag{1.1}$$

The one-dimensional transformation as given above relates the coordinates of any event as noted in the frame $S$ in terms of those as recorded in the frame $S'$. In Newtonian mechanics there is a single universal time $t$ and the last equation of (1.1) expresses precisely this, namely, the absolute character of time—a feature that is independent of the relative motion of the two velocities.

Differentiating by $t$ leads to

$$U' = U + u \tag{1.2}$$

where $U = \frac{dx}{dt}$ and $U' = \frac{dx'}{dt}$. The above equation represents the classical velocity-addition formula.

We can generalize (1.1) to three dimensions for two systems labeled by the coordinates $\vec{r} \equiv (x, y, z)$ in the frame $S$ and $\vec{r'} \equiv (x', y', z')$ in the frame $S'$ where the latter is moving uniformly with velocity $\vec{V} \equiv (u, v, w)$ in relation to S. To this end, we write down

$$\vec{r'} = \vec{r} + \vec{V}t \Rightarrow x' = x + ut, \quad y' = y + vt, \quad z' = z + wt \tag{1.3}$$

with $t' = t$ meaning time to be absolute in that it is independent of the relative motion of the two observers in $S$ and $S'$. In Equation (1.3), (u,v,w) are the components of the constant velocity $\vec{V}$. Such transformations are the standard or pure Galilean transformations and the motion is said to be invariant under them.

It is possible further to extend (1.3) by the inclusion of a static rotation and an arbitrary spatial shift $\vec{\alpha}$, i.e.,

$$\vec{r'} = R\vec{r} + \vec{V}t + \vec{\alpha}, \quad \vec{\alpha} \in \mathbb{R}^3 \tag{1.4}$$

along with a constant temporal translation $\xi$, i.e.,

$$t' = t + \xi, \quad \xi \in \mathbb{R} \tag{1.5}$$

where $R$ stands for a $3 \times 3$ orthogonal rotation matrix and is given by

$$R = \begin{pmatrix} \cos\gamma\cos\beta\cos\alpha - \sin\gamma\sin\alpha & \cos\gamma\cos\beta\sin\alpha + \sin\gamma\cos\alpha & -\cos\gamma\sin\beta \\ -\sin\gamma\cos\beta\cos\alpha - \cos\gamma\sin\alpha & -\sin\gamma\cos\beta\sin\alpha + \cos\gamma\cos\alpha & \sin\gamma\sin\beta \\ \sin\beta\cos\alpha & \sin\beta\sin\alpha & \cos\beta \end{pmatrix} \tag{1.6}$$

where $\alpha$, $\beta$ and $\gamma$ are the three Eulerian angles of rotations. The domains of these angles are $0 \leq \alpha < 2\pi$, $0 \leq \beta \leq \pi$ and $0 \leq \gamma < 2\pi$. The rotation matrix $R$ satisfies $|R| = 1$ and $RR^T = R^T R = I$.

Evidently the generalized Galilean transformations (1.4) are a set of linear equations. Given an inertial frame $S$, it can be carried over to another inertial frame $S'$ in 10 possible ways. These correspond to three for spatial translations, three for rotations, one for time translation and three for boosts denoted by the constant velocity $\vec{V}$ such as in (1.3). All these constitute a 10-parameter Galilean group. Representing an element of such a group as $h(R, \vec{V}, \vec{\alpha}, \xi)$, the composition rule is defined as

$$h_3(R_3, w, \alpha_3, \xi_3) = h_2(R_2, v, \alpha_2, \xi_2) \circ h_1(R_1, u, \alpha_1, \xi_1) \tag{1.7}$$

where the various entries stand for

$$R_3 = R_2 R_1, \qquad w = v + R_2 u, \quad \alpha_3 = \alpha_2 + R_2\alpha_1 + v\xi_1, \quad \xi_3 = \xi_1 + \xi_2 \tag{1.8}$$

In other words, successive operations of two rotations is equivalent to a single rotation and reveals the essential feature that makes a group.

Newton's second law considers the momentum variable to be fundamental and is concerned with the rate of change in it when an external force is applied. Operationally the force acting upon an object of constant mass $m$ is given by a quantity which is the product of $m$ times its acceleration $\vec{a}$:

$$\vec{F}(\vec{r}, \dot{\vec{r}}, t) = m\ddot{\vec{r}} = m\vec{a} \tag{1.9}$$

where $\vec{a} = \ddot{\vec{r}}$ is the acceleration and the force function is assumed to depend on the position, velocity and time. Thus, an application of a force causes the body to accelerate. Without it the body stays at rest and the motion is restricted by the equation $m\ddot{\vec{r}} = 0$. $m$ is called the inertial mass: it is the ability of the body to resist being accelerated. With $\vec{F}$ known, a larger $m$ points to a smaller $\vec{a}$.

At the heart of the second law lies the postulate that the forces are linearly additive and behave as vectors. So, in an interacting system, if $n$ forces $\vec{F_1}, \vec{F_2}, ...\vec{F_n}$ to act on a body of mass $m$, it produces an acceleration

$$\vec{a} = \frac{1}{m}(\vec{F_1} + \vec{F_2} + ... + \vec{F_n}) = \frac{\vec{F}}{m} \tag{1.10}$$

(1.10) is an extended version of Equation (1.9) and embodies the central idea of the *principle of linear superposition*. The main point of the second law is that an application of a force brings about a change in the velocity of the object which in turn causes it to accelerate. Conversely, a time-change in velocity accounts for the force. Knowing the force function $\vec{F}$ along with the initial conditions $x(t) = x_0$ and $\dot{x}(t) = \dot{x}_0$ at time $t = t_0$ allows one to draw a conclusion about the trajectory of the particle that the particle actually traces out.

Classical mechanics is characterized by two underlying principles governing it, namely:

(i) Newton–Laplace principle of determinacy (or the deterministic law) which states that the state of the system (comprising the position and velocity of the particle) at time $t$ completely determines its behavior for all future ($> t$) and past ($< t$) times; in other words, the laws of physical phenomena are entirely deterministic and

(ii) Galilean principle of relativity which states that the law of motion (1.9) has the same form in every inertial frame.

Suppose that a physical state of a mechanical system is known at the initial time $t_0$ as given by the coordinate $\vec{r}(t_0) = \vec{r_0}$ and velocity $\dot{\vec{r}}(t_0) = \dot{\vec{r_0}} = \vec{v_0}$. The principle of determinacy tells us that the motion given by the function $\vec{r} = \vec{r}(t, \vec{r_0}, \dot{\vec{r_0}})$ is uniquely known for all $t > t_0$ and $t < t_0$. Assuming such a function to be sufficiently smooth a couple of differentiations with respect to $t$ yields the form $\ddot{\vec{r}}(t_0) = \vec{f}(t_0, \vec{r_0}, \dot{\vec{r_0}})$ in which $t = t_0$ has been set. Since $t_0$ can be chosen arbitrarily, the equation

$$\ddot{\vec{r}}(t) = m\vec{f}(t, \vec{r}, \dot{\vec{r}}) = \vec{F}(t, \vec{r}, \dot{\vec{r}}) \tag{1.11}$$

is the one that describes the motion of a particle of mass $m$ and is valid for all $t$. Notice that in writing down the above differential equation we have not specified any frame of reference. If the frame of reference is identified with an inertial frame, then (1.11) is identical with Newton's equation of motion (1.9) with $\vec{F}$ as the external force acting on the particle with mass $m$.

Conversely, given (1.11) and the initial conditions $\vec{r} = \vec{r_0}, \dot{\vec{r}} = \dot{\vec{r_0}} = \vec{v_0}$ at time $t = t_0$ with the function $f$ assumed to be sufficiently smooth we know that there exists a unique solution in the form $\vec{r} = \vec{r}(t, \vec{r_0}, \dot{\vec{r_0}})$. This implies the principle of determinacy. Note that the kind of law by which the system is deterministic in only one direction (say, for instance, the future) but not in the other (the past) or conversely is forbidden.

Turning to Newton's third law, it states that if a body exerts a force on another object then the latter returns the force with an amount equal and opposite to what it receives. In the case of the rigid body, the forces between the constituent particles that make up the rigid body are then equal and opposite. These are the internal forces and do not contribute to producing any effect on the motion of the body. Taken with the first law, we can say that the deviation from the state of rest or uniform motion is caused only by the external forces.

Consider the specific case of the component $\vec{F_i}$ in (1.10) when the $i^{th}$ particle exerts a force on the body of mass $m$. By Newton's third law, it has to be equal and opposite to what $m$ exerts on the $i^{th}$ particle but it is tacitly assumed that other particles (excluding the $i^{th}$ particle) in the system are just spectators and their role in the contribution to this mutual force-play is taken to be too insignificant to affect the linearity of the relation (1.10). The principle of linear superposition disregards all kinds of self-interactions and other many-body interactions.

We now move to an elementary consequence. Consider three particles $i$, $j$ and $k$ with respective masses $m_i$, $m_j$ and $m_k$ existing in an isolated environment. Labeling $\vec{F^{ij}}$ and $\vec{F^{kj}}$ as, respectively, the force that the $i^{th}$ particle exerts on the $j^{th}$ particle and the one that the $k^{th}$ particle exerts on the $j^{th}$ particle, we can write for the $j^{th}$ particle

$$m^j \vec{a^j} = \vec{F^{ij}} + \vec{F^{kj}} \tag{1.12}$$

Similarly for the $k^{th}$ particle, which would be influenced by the $i^{th}$ and $j^{th}$ particles, we would have

$$m^k \vec{a^k} = \vec{F^{ik}} + \vec{F^{jk}} \tag{1.13}$$

By Newton's third law the forces $\vec{F^{kj}}$ and $\vec{F^{jk}}$ are equal and opposite and so when we add (1.12) and (1.13) the effects of these forces cancel leaving us with the relation

$$m^j \vec{a^j} + m^k \vec{a^k} = \vec{F^{ij}} + \vec{F^{ik}} \tag{1.14}$$

If the accelerations $\vec{a}_j$ and $\vec{a}_k$ are now taken to be the same, say equal to $\vec{a}$, then (1.14) points to an equation of motion of a single mass $M$ according to the form

$$M\vec{a} = \vec{F^{ij}} + \vec{F^{ik}} \tag{1.15}$$

where $M = (m^j + m^k)$ and the right-hand side expresses the total force that the $i^{th}$ particle exerts on the $j^{th}$ and $k^{th}$ particles. The above result is easily generalizable to the case of $N$ interacting particles.

## 1.2 Homogeneity and isotropy

An important class of invariance in the Newtonian theory is the one that is due to the time translation. This means that if (1.9) admits of a solution of the type $\vec{r} = \vec{f}(t)$, then the class denoted by $\vec{r} = \vec{f}(t + \xi)$ also stands for a solution for $\xi \in \mathbb{R}$. In other words, (1.9) depicts an autonomous character, i.e., there can be no explicit time-dependence in $\vec{F}$:

$$m\ddot{\vec{r}} = \vec{F}(\vec{r}, \dot{\vec{r}}) \tag{1.16}$$

The implication is that every point of time is as good as any other moment.

Apart from the time translation, another class of invariance pertains to space-translations in $\mathbb{R}^3$. A consequence is that if (1.10) takes in a solution of the type $\vec{r_i} = \vec{f_i}(t), \quad i = 1, 2, ...n$, corresponding to the motion of a system of $n$ particles, then $\vec{r_i} = \vec{f_i}(t) + \vec{\alpha_i}, \quad i = 1, 2, ...n$, where $\vec{\alpha_i}$ is a constant vector, also passes for an equally acceptable class of solutions. The message is that every point in the Universe is equivalent to every other point. Qualitatively the concept of homogeneity implies that the properties of the body do not change or vary from one point in the body to another. Indeed on sufficient large scales, Newtonian space and time are both homogeneous and isotropic.

Homogenity in space also implies that the force function $\vec{F_i}$ on the $i^{th}$ particle in an inertial coordinate system can be only a function of the difference of two coordinates like $\vec{r_j} - \vec{r_i}$ and the relative velocities $\dot{\vec{r_j}} - \dot{\vec{r_i}}$. Thus

$$m_i\ddot{\vec{r_i}} = \vec{F_i}(\vec{r_j} - \vec{r_i}, \dot{\vec{r_j}} - \dot{\vec{r_i}}) \tag{1.17}$$

where $i, j = 1, 2, ...n$.

Newtonian space is additionally isotropic implying that there is no preferential direction. To put it simply, if R represents an orthogonal transformation in the three-dimensional Euclidean space $\mathbb{R}^3$, then isotropic invariance expresses invariance under rotations as given by

$$\vec{F}(R\vec{r}, R\dot{\vec{r}}) = R\vec{F}(\vec{r}, \dot{\vec{r}}) \tag{1.18}$$

where $R$ is an orthogonal matrix operating on each component of $\vec{r}$. Of course homogeneity suggests isotropy but the opposite is not necessarily true. Consider a smooth flat region with a sea of uniformly green all around and one is occupying a position in an elevated site somewhere in the middle of it. Whichever way you look it will appear the same in all directions. We say that the space is isotropic. However, because of one's privileged location one cannot take for granted the space to be homogeneous as well. For a homogeneous region one point in it can be substituted for another and so it is left to our choice as to which one is preferred irrespective of which direction it is located.

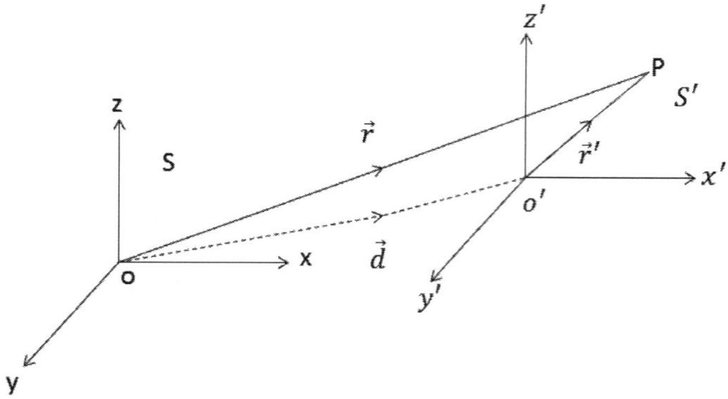


**FIGURE 1.1**: Two coordinate frames $S$ and $S'$ which are in translatory motion with respect to each other. $P$ is the representative point whose coordinates with respect to the origins $O$ and $O'$ are, respectively, $\vec{r}$ and $\vec{r'}$.

Returning to the frames $S$ and $S'$ discussed previously, let us think of a translatory motion of $S'$ with respect to $S$. Such a motion does not change the directions of the three spatial axes of $S'$ relative to the corresponding ones of $S$. In other words, they remain parallel to one another.

Let us consider now the motion of a particle at a point $P$ whose coordinates with respect to the origins $O$ and $O'$ are, respectively, $\vec{r}$ and $\vec{r'}$. See the configuration in Figure 1.1 in which $|\vec{d}|$ is the distance between $O$ and $O'$. It follows from the vector addition formula namely, $\vec{r} = \vec{d} + \vec{r'}$, that in addition to velocity addition rule (1.2) a similar one for the acceleration also holds if we further differentiate it with respect to t. It follows that

$$\vec{a} = \ddot{\vec{d}} + \vec{a'} \tag{1.19}$$

where $\ddot{\vec{d}}$ could be interpreted as the acceleration of $S'$ with respect to $S$ while $\vec{v'}$ and $\vec{a'}$ stand, respectively, for the velocity and acceleration of the particle at $P$ with respect to $S'$.

For a non-accelerating frame $S'$, clearly $\ddot{\vec{d}}$ vanishes, implying from (1.19) that the two accelerations coincide: $\vec{a} = \vec{a'}$. Consequently the law of motion is similar in the two frames $S$ and $S'$, i.e., $\vec{F} = m\vec{a} = m\vec{a'}$, reflecting the inertial character of Newtonian mechanics. On the other hand, for an accelerating moving frame the term $\ddot{\vec{d}}$ cannot vanish and we have a modified form of the second law of motion $\vec{F'} = m\vec{a'}$ where the modified force $\vec{F'}$ is defined by $\vec{F'} = \vec{F} - m\ddot{\vec{d}}$. In the latter form, the additional term -m$\ddot{\vec{d}}$ arising from the accelerating character of the coordinate frame $S'$ is generally responsible for the occurrence of a fictitious force term.

## 1.3 Solution process

Determining the solutions for a given physical system involves basically two fundamental steps. The first is to choose an appropriate reference frame with respect to which the given problem is defined unambiguously while the other is to set up the differential equation as guided by Newton's second law.

Concerning the first step, the Cartesian system of coordinates is often the simplest reference frame to work with. If $F_x$, $F_y$, $F_z$ are the components of an external force $\vec{F}$ along the $x, y, z$ directions, respectively, then for a particle of constant mass $m$, we have according to Newton's second law, the component equations, $F_x = m\ddot{x}, F_y = m\ddot{y}, F_z = m\ddot{z}$.

For an $N$-point particle system, we can similarly write down relations of the type

$$m_1\ddot{x}_1 = F_{x_1}, \quad m_2\ddot{x}_2 = F_{x_2}, \quad ... \quad m_N\ddot{x}_N = F_{x_N}, \tag{1.20}$$

$$m_1\ddot{y}_1 = F_{y_1}, \quad m_2\ddot{y}_2 = F_{y_2}, \quad ... \quad m_N\ddot{y}_N = F_{y_N}, \tag{1.21}$$

$$m_1\ddot{z}_1 = F_{z_1}, \quad m_2\ddot{z}_2 = F_{z_2}, \quad ... \quad m_N\ddot{z}_N = F_{z_N}, \tag{1.22}$$

In more compact terms we have

$$m_i\ddot{x}_i = F_{x_i}, \quad m_i\ddot{y}_i = F_{y_i}, \quad m_i\ddot{z}_i = F_{z_i} \quad i = 1, 2, ..., N \tag{1.23}$$

It should be mentioned that when we switch over to an arbitrary reference frame the transparency of the force-acceleration relationships, as provided by the Cartesian form, may be lost. As a specific example, we may think of the two-dimensional plane polar coordinates $(r, \theta)$ system. Here the forms of the velocity and acceleration are

$$\vec{v} = \dot{r}\hat{r} + r\dot{\theta}\hat{\theta} \tag{1.24}$$

$$\vec{a} = (\ddot{r} - r\dot{\theta}^2)\hat{r} + (r\ddot{\theta} + 2\dot{r}\dot{\theta})\hat{\theta} \tag{1.25}$$

where $\hat{r}, \hat{\theta}$ are unit vectors along the directions of $r$ and $\theta$. We see that $F_r \neq m\ddot{r}$ and $F_\theta \neq m\ddot{\theta}$, $F_r$ and $F_\theta$ being, respectively, the components of the external force in the $r$ and $\theta$ directions.

Next, let us probe into the interplay of (1.24) and (1.25). Consider the simple case of a particle moving uniformly along a circle with a constant angular velocity $\omega = \dot{\theta}$. Then the acceleration turns out to be simply $\vec{a} =$ -$\omega^2\ \vec{r}$. The negative sign implies that the acceleration is directed *toward* the origin with a constant magnitude. The occurrence of acceleration is due to the fact that the direction of the velocity vector as given by (1.24) is continuously changing at a constant rate. Such a kind of acceleration is referred to as the *centripetal* acceleration.

Concerning the use of other reference frames, there are indeed situations depending on the type of problem at hand, for instance the central force problem, where a great deal of simplification results if the plane polar coordinates, $(r, \theta)$, are employed. Some well-known examples of the central force are the inverse Newton's square law of gravitation and Coulomb's electrostatic force between two charges.

Apart from the Cartesian reference frame, use of a spherical polar system having coordinates $(r, \theta, \phi)$ or a cylindrical system having coordinates $(\zeta, \Phi, z)$ can be convenient. The relations in terms of $(x, y, z)$ and the corresponding Laplacian $\nabla^2$ are given by

$$x = r\sin\theta\cos\phi, \quad y = r\sin\theta\sin\phi, \quad z = r\cos\theta \tag{1.26}$$

$$\nabla^2 = \frac{1}{r^2}\frac{\partial}{\partial r}(r^2\frac{\partial}{\partial r}) + \frac{1}{r^2\sin\theta}\frac{\partial}{\partial\theta}(\sin\theta\frac{\partial}{\partial\theta}) + \frac{1}{r^2\sin^2\theta}\frac{\partial^2}{\partial\phi^2} \tag{1.27}$$

and

$$x = \zeta\cos\Phi, \quad y = \zeta\sin\Phi, \quad z = z \tag{1.28}$$

$$\nabla^2 = \frac{1}{\zeta}\frac{\partial}{\partial\zeta}(\zeta\frac{\partial}{\partial\zeta}) + \frac{1}{\zeta^2}\frac{\partial^2}{\partial\Phi^2} + \frac{\partial^2}{\partial z^2} \tag{1.29}$$

There also exist other coordinate frames such as the parabolic system $(\zeta, \eta, \varphi)$ wherein the correspondences are defined by

$$x = \sqrt{\zeta\eta}\cos\varphi, y = \sqrt{\zeta\eta}\sin\varphi, z = \frac{1}{2}(\xi - \eta). \tag{1.30}$$

$$\nabla^2 = \frac{4}{\xi+\eta}\frac{\partial}{\partial\xi}(\xi\frac{\partial}{\partial\xi}) + \frac{4}{\xi+\eta}\frac{\partial}{\partial\eta}(\eta\frac{\partial}{\partial\eta}) + \frac{1}{\xi\eta}\frac{\partial^2}{\partial\phi^2} \tag{1.31}$$

Proceeding now to the second step, a differential equation that follows from (1.9) has to be formulated. In this regard, we have to identify the guiding forces and make sure that the conditions prescribed in the given problem are the appropriate ones for which Newton's law is applicable. In general if the constraints are present, which are so ordinarily, then the external forces are not known completely. To tackle this issue we have to go for an analytical formalism that will be explained in Chapter 3.

We have already referred to the consideration of the inertial frame. Others are

(i) The magnitude of the masses and time-distance scales should neither be too small (like the dimensions we deal with at the microscopic level of atoms and nuclei where the principles of quantum mechanics are applicable) nor too large (as we encounter in the solar system or a galaxy wherein the underlying dynamics are governed by the more sophisticated Einstein's theory of general relativity) and

(ii) The magnitude of the velocity of the particle must be very small compared to the velocity of light $c$. It is well known that dynamics of objects moving with high velocities, comparable to the order of $c$, come under the purview of the formulae of the special theory of relativity. In the latter formalism, Galilean transformations are replaced by the Lorentz transformations that do not hold the time variable $t$ to be absolute but is treated at par with the space coordinates.

---

## 1.4 Conservative forces

The conservative force represents a special class of force such that the work done by it, as the system evolves from one configuration to another, depends on its initial and final positions only. Its occurrence in nature is quite commonplace. Gravity and elastic spring force are the two familiar examples of conservative force. Friction, however, is a non-conservative force. A conservative force can be distinguished by any one of the following features:

(i) The work done by the force is path independent. In other words, no matter what route is chosen, the work done by the force acting upon the particle will be the same as long as it is conservative.

(ii)Around a closed path or loop in a simply connected region the work done by the force is zero: $\oint \vec{F}.d\vec{r} = 0$ for any closed loop $C$.

(iii) If $(F_x, F_y, F_z)$ are the Cartesian components of the force $\vec{F}$, then the sum $F_x dx + F_y + F_z dz$ is an exact differential.

(iv) $\vec{F}$ is only a function of position and $\nabla$ x $\vec{F} = 0$.

(v) A potential energy function $V$ exists that has a definite value at every point.

(vi) $T + V$ = constant where $T$ is the kinetic energy.

From (iv) we can write $\vec{F}$ as the gradient of some scalar function $V$. It also points to the freedom that an arbitrary constant can be added to $V$ without altering $\vec{F}$. Since the gradient points to the direction of increasing potential and forces cause the system to move to a lower potential, a negative sign is chosen to express $\vec{F} = -\nabla V$: in other words, the force is the negative gradient of some potential function $V$. In one dimension it is simply $F_x = -\frac{dV}{dx}$. When integrated it gives immediately for the work done, $W = -[V(b) - V(a)] = -\triangle V$. We are therefore led to the following principle that the work done on the system (positive work) increases the potential energy while the work done by the system (negative work) decreases the potential energy.

Sometimes a force can be a combination of both conservative and non-conservative parts. Then, of course, the work done by the non-conservative part will only contribute to the change in energy as the motion is executed between two different points. Let us now give some examples of conservative and non-conservative forces.

**Example 1.1**

The force of gravity between two objects, of masses M and m and $|\vec{r}|$ as the separating distance between them, is given by

$$\vec{F^i} = -\frac{GMm}{r^3}\vec{r}, \quad r \neq 0$$

Around a closed path

$$\oint -\frac{GMm}{r^3}\vec{r}.d\vec{r} = \oint -\frac{GMm}{r^2}dr = \Delta\frac{GMm}{r} = 0, \quad r \neq 0$$

Thus gravity is a conservative force.

**Example 1.2**

The frictional force, given an infinitesimal displacement $d\vec{r}$, always acts oppositely: $\vec{F}.d\vec{r} < 0$. As a result $\oint \vec{F}.d\vec{r} < 0$. Therefore, friction is a non-conservative force.

**Example 1.3**

Consider a force $\vec{F}$ with components $(y, -x, z)$ and $C$ to be a circle defined by $x^2 + y^2 = c^2$, $z = h$. With the parametrization $x = c\cos\theta$ and $y = c\sin\theta$ for the circle, it is easy to work out

$$\int_0^{2\pi} (c\sin\theta, -c\cos\theta, h).(-c\sin\theta, c\cos\theta)d\theta = -2\pi c^2 \neq 0$$

So the force $\vec{F}$ here is a non-conservative one.

---

## 1.5 Simple harmonic motion

The simple harmonic motion (SHM) is perhaps the simplest and most elegant of all dynamical systems. It considers the problem of a mass-spring system being attracted to a given fixed point by a force which by *Hooke's law* is assumed to be directly proportional to the distance from the point. Obviously the SHM, which is also referred to as the simple harmonic oscillator, is a conservative system. In a one-dimensional setting, along the x-axis, the force is given by $F(x) = -kx$, where $k > 0$, which is linear and of a restoring type. $F(x)$ being an odd function of $x$ is negative when $x > 0$ and positive when $x < 0$. For such a force we run into the differential equation as given by Newton's second law

$$m\ddot{x} + kx = 0 \quad \rightarrow \ddot{x} + \omega_0^2 x = 0, \quad \omega_0 = \sqrt{\frac{k}{m}} \tag{1.32}$$

where the overhead dots represent derivatives with respect to the time $t$ and $\omega_0$ is the *natural* frequency (angular) having a dimension $(\text{time})^{-1}$. A system whose dynamics is governed by such an equation is called a harmonic oscillator.

Equation (1.32) admits the following trigonometric solution:

$$x(t) = A\cos(\omega_0 t + \lambda) \tag{1.33}$$

The coefficient $A$ is the amplitude of the motion and $\lambda$ is the initial phase at $t = 0$. To get an estimate of the period of the motion T, the solution being of a trigonometric type, we have to consider the repeated intervals of time for which the motion is similar due to the cosine nature of the solution. The value of T is obtained by noting that each time the phase angle changes by $2\pi$, both $x$ and the velocity $v = \frac{dx}{dt}$ undergo a complete cycle of variable. In this way corresponding to the pairs, $(\phi_1, t_1)$ and $(\phi_1 + 2\pi, t + T)$, we have

$$\phi_1 = \omega_0 t_1 + \lambda, \quad \phi_1 + 2\pi = \omega_0(t_1 + \mathrm{T}) + \lambda \tag{1.34}$$

By subtraction we get $\omega_0 \mathrm{T} = 2\pi$ which yields

$$\mathrm{T} = \frac{2\pi}{\omega_0} = 2\pi\sqrt{\frac{k}{m}} \tag{1.35}$$

The inverse of the period $T$ gives the frequency $\nu$ measured in Hertz, i.e., cycles per second: $\nu = \frac{1}{\mathrm{T}}$. Because of (1.35) it is clear that the period or frequency is independent of the amplitude. This is a very important property as far as the simple harmonic motion is concerned. Such a system of bounded periodic motion is referred to as an isochronous system. In the literature only a limited number of isochronous systems are known.

Having a general solution (1.33) involving two arbitrary constants $A$ and $\lambda$, we need certain initial conditions to fix them. For instance, the initial conditions may correspond to the initial position $x_0$ and initial velocity $v_0$ of the particle at $t = 0$. Since

$$v = \frac{dx}{dt} = -\omega_0 A\sin(\omega_0 t + \lambda) \tag{1.36}$$

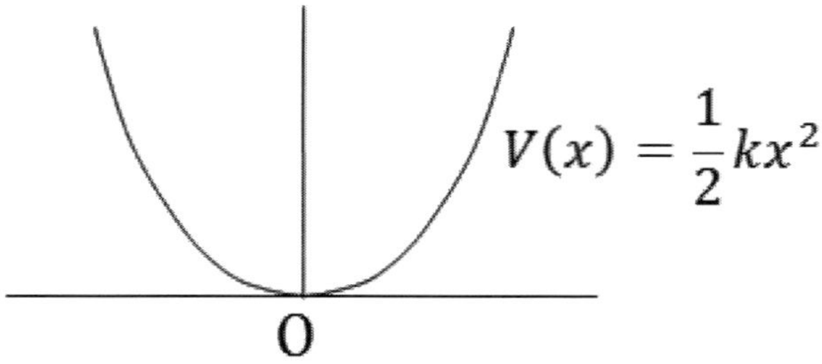


**FIGURE 1.2**: The parabolic form of the potential $V(x) = \frac{1}{2}kx^2$.

it turns out that

$$x_0 = A\cos\lambda, \quad v_0 = -\omega_0 A\sin\lambda \tag{1.37}$$

The quantities $A$ and $\lambda$ can now be solved to get

$$A = \sqrt{(x_0)^2 + \frac{v_0^2}{\omega_0^2}}, \quad \lambda = -\arctan(\frac{v_0}{\omega_0 x_0}) \tag{1.38}$$

We now make a few remarks on the character of the potential function $V(x)$ associated with the SHM. We say that the particle is in an equilibrium (or stationary) state if it is not subject to any force. This implies $\frac{dV}{dx} = 0$ and we observe that an equilibrium state of the particle pertains to being a minimum or maximum of the potential. To make the subtle distinction between when the equilibrium is stable or unstable we have to check whether forcing small deviations from the equilibrium point $x = x_0$ causes the particle to return to the point. Mathematically, this means that the requirement of stable equilibrium corresponds to the inequality $\frac{dF}{dx} < 0$ at $x = x_0$: in other words, if the particle steps to the right then the force acts to the left while the opposite is true when the particle is displaced to the left. On the other hand, if $\frac{dF}{dx} > 0$ the contrary features hold and we run into an unstable equilibrium state. Hence because $F = -\frac{dV}{dx}$, a stable (unstable) equilibrium point signals the inequality $\frac{d^2V}{dx^2} > (<)0$ and so corresponds to the minimum (maximum) of the potential function $V(x)$. The third possibility $\frac{dV}{dx} = 0 = \frac{d^2V}{dx^2}$ points to a neutral equilibrium point.

For the simple harmonic oscillator the force term implies that the potential is given by $V(x) = \frac{1}{2}kx^2$ and so the total mechanical energy of the system, being the sum of of kinetic energy and potential energy, is

$$E = \frac{1}{2}mv^2 + V(x) = \frac{1}{2}m(\frac{dx}{dt})^2 + \frac{1}{2}kx^2 \tag{1.39}$$

The positivity of the kinetic energy restricts $E > 0$ implying $E$ needs to be greater than $V$. For $E = 0$, the particle is at rest at the lowermost point

$O$ of the parabola as shown in Figure 1.2. It is always stable there: small displacements from $O$ inevitably cause the particle to return to $O$. For $E > 0$, however, the inequality $V(x) \leq E$ has to hold, i.e., the particle remains within the confines of the parabola: $E \leq \frac{1}{2}kx^2$. The roots of the latter give the turning points, $a = -\sqrt{\frac{2E}{k}}$ and $b = \sqrt{\frac{2E}{k}}$. The regions beyond these turning points are classically inaccessible.

Consider a smooth and continuous function $V(x)$. Taylor series expansion about an arbitrary point $x = x_0$ gives

$$V(x) = V(x_0) + (x - x_0)V'(x_0) + \frac{1}{2}(x - x_0)^2 V''(x_0) + \cdots + \frac{1}{n!}(x - x_0)^n V^{(n)}(x_0) + R_{n+1} \tag{1.40}$$

where $R_{n+1}$ corresponds to the remainder after $(n + 1)$ terms.

If $x_0$ is an equilibrium point, i.e., $V'(x_0) = 0$ and if we retain terms up to $O((x - x_0)^2)$, then for small displacements from the equilibrium

$$V(x) \approx V(x_0) + \frac{1}{2}(x - x_0)^2 V''(x_0) \tag{1.41}$$

At the point $O$ where the potential has the minimum, $V''(x_0) > 0$. Should we redefine the zero to set $V(x_0) = 0$, the potential $V(x)$ simplifies to

$$V(x) \sim \frac{1}{2}(x - x_0)^2 V''(x_0) \tag{1.42}$$

which has the profile of a harmonic oscillator potential having an angular frequency given by $\sqrt{\frac{V''(x_0)}{m}}$. From such a general consideration, we therefore conclude that any physical system near its stable equilibrium point can always be approximated by a harmonic oscillator potential. In fact, for many physical phenomena, whenever we consider the motion of a particle being subjected to a potential that has one or more local minima, the harmonic oscillator approximation gives the initial clue to the understanding of the behavior of the system.

---

## 1.6 Damped and forced oscillator

### 1.6.1 Damped oscillator

In reality, oscillations tend to get overpowered by frictional forces that have the character to reduce or dampen the amplitude causing the system to come ultimately to a halt. A damped oscillator is a more complicated case compared to the simple harmonic motion.

In the presence of a frictional term which is proportional to the velocity of motion, i.e., the case of viscous damping, (1.32) can be extended to the form $m\ddot{x} + kx + \eta\dot{x} = 0$. The parameter $\eta$ is referred to as the damping constant. In the following we reexpress this equation in terms of a quantity $\lambda$

$$\ddot{x} + \omega_0^2 x + \lambda\omega_0\dot{x} = 0 \tag{1.43}$$

where we have put $\eta = m\lambda\omega_0$ relating the damping factor with the natural frequency parameter $\omega_0$.

The linearity of Equation (1.43) suggests that we can try a solution of the type $x(t) \sim e^{\alpha t}$. Substitution in (1.43) implies that $\alpha$ should be restricted by

$$\alpha = -\frac{\lambda\omega_0}{2} \pm \delta, \quad \delta = \omega_0\sqrt{\frac{\lambda^2}{4} - 1} \tag{1.44}$$

A look at the form for $\delta$ reveals that three different cases arise according to whether (i) $\lambda < 2$, (ii) $\lambda = 2$ or (iii) $\lambda > 2$. Let us consider these cases separately.

Case (i): $\lambda < 2$

Here $\delta$ is imaginary and the general solution of (1.43) can be written as

$$x = Ae^{\frac{-\lambda\omega_0 t}{2}} \cos[(\omega_0\sqrt{1 - \frac{\lambda^2}{4}}t + \phi] \tag{1.45}$$

where $A$ and $\phi$ are integration constants. An appropriate choice of the initial conditions enables one to fix these constants. An imaginary $\delta$ corresponds to the case of underdamping. The solution (1.45) represents an oscillation of amplitude $Ae^{-\frac{\lambda\omega_0 t}{2}}$ with a frequency $\bar{\omega}$ as given by $\bar{\omega} \equiv \omega_0\sqrt{1 - \frac{\lambda^2}{4}}$. The amplitude can be seen to decay exponentially with time while the frequency $\bar{\omega}$ is lower than the natural frequence $\omega_0$ of the standard harmonic oscillator. Actually Equation (1.43) can be converted to the form $\ddot{z} + \bar{\omega_0}^2 z = 0$ by making use of the transformation $x \to Be^{-\frac{\lambda\omega_0 t}{2}} z$. This is an SHM but the oscillation happens with a reduced or damped frequency $\bar{\omega}$. The process is carried on until eventually the equilibrium state is reached.

Case (ii): $\lambda = 2$

This case corresponds to $\delta = 0$. With the general solution given by

$$x = x_0 e^{\frac{-\lambda\omega_0 t}{2}} \tag{1.46}$$

we have here a critical damping case. Note that the system decays exponentially to the equilibrium position.

Case (iii): $\lambda > 2$

Such a possibility for $\lambda$ implies $\delta$ to be real and the system is overdamped. Equation (1.43) admits the following class of general solution:

$$x = e^{\frac{-\lambda\omega_0 t}{2}}(Ce^{\delta t} + De^{-\delta t}) \tag{1.47}$$

where C and D are constants of integration to be determined from the initial conditions. Clearly of the two terms in the right-hand side, one is smaller in magnitude in comparison to the other and gets to a steady state at an earlier time. The point to note is that it is achieved slower than the critically damped case.

### 1.6.2 Forced oscillator

#### (a) Forced harmonic oscillator

Let us next consider the case of a forced harmonic oscillator subject to a time-varying periodic external force $F(t)$ given by $F(t) = F_0 \cos(\omega t)$ on the undamped motion of the simple harmonic oscillator:

$$\ddot{x} + \omega_0^2 x = \frac{F_0}{m}\cos(\omega t) \tag{1.48}$$

where $\omega$ denotes the *driving* frequency. The general solution of (1.48) can be expressed in the form

$$x(t) = A\cos(\omega_0 t - \phi) + \bar{A}\cos(\omega t), \quad \bar{A} = \frac{F_0/m}{(\omega_0^2 - \omega^2)} \tag{1.49}$$

where the first term in the right-hand side corresponds to a solution of the simple harmonic motion, $\omega_0$ being its *natural* frequency. The solution (1.49) is a composition of two cosine-like terms valid at different frequencies $\omega_0$ and $\omega$. We need to distinguish between the two cases when $\omega \neq \omega_0$ and $\omega = \omega_0$.

The case $\omega \neq \omega_0$ is a simple one. We notice that the second amplitude $\bar{A}$ in (1.49) has a finite value when $\omega \to 0$ while it approaches zero as $\omega \to \infty$. However, the situation changes drastically when $\omega$ approaches $\omega_0$. Here we have to replace the solution (1.49) by a different form

$$x(t) = P\cos(\omega_0 t) + \frac{F_0}{2m\omega_0} t \sin(\omega_0 t) \tag{1.50}$$

To arrive at (1.50) we took the representation $x(t) = P\cos(\omega_0 t) + Qt\sin(\omega_0 t)$ as a plausible solution of (1.48) and fixed the arbitrary constants $P$ and $Q$ through matching of the two sides. We notice that the second term in the right-hand side of (1.50) blows up as $t \to \infty$. Also, because of the presence of the sine term it oscillates between -$\frac{F_0}{2m\omega_0}$ and $\frac{F_0}{2m\omega_0}$ and dominates the oscillation of the first term as $t$ grows. Thus by driving the system to a certain frequency value we are met with a large oscillation behavior. Such a feature is referred to as the *resonance.*

### (b) Forced harmonic oscillator with linear damping

What happens if a frictional term is added in (1.48)? We will then have to deal with an explicitly velocity-dependent differential equation with a time-varying source term:

$$\ddot{x} + \omega_0^2 x + \lambda\omega_0 \dot{x} = \frac{F_o}{m}\cos(\omega t) \tag{1.51}$$

To handle such an equation the usual trick is to complexify it and identify it as its real part. Toward this end we introduce the complex variable $z = x + iy$ and notice that Equation (1.51) along with its partner equation containing a sine-like external force term, namely,

$$\ddot{y} + \omega_0^2 y + \lambda\omega_0 \dot{y} = \frac{F_o}{m}\sin(\omega t) \tag{1.52}$$

are the respective real and imaginary components of the following single equation:

$$\ddot{z} + \omega_0^2 z + \lambda\omega_0 \dot{z} = \frac{F_o}{m}e^{i\omega t} \tag{1.53}$$

where we have made use of the trigonometric identity

$$e^{i(\omega t)} = \cos(\omega t) + i\sin(\omega t) \tag{1.54}$$

to arrive at the right-hand side of (1.53).

We now seek a particular solution of (1.53) as

$$z = z_0 e^{i\omega t} \tag{1.55}$$

We immediately notice that such a solution is perfectly tenable provided the following condition is fulfilled:

$$z_0 = \frac{F_o}{m}\frac{1}{[-\omega^2 + i\lambda\omega\omega_0 + \omega_0^2]} \tag{1.56}$$

The modulus value of $z_0$ is recognized as

$$|z_0| = \frac{F_o}{m}\frac{1}{[(-\omega^2 + \omega_0^2)^2 + \lambda^2\omega^2\omega_0^2]^{\frac{1}{2}}} \tag{1.57}$$

signalling the proportionality of the amplitude $|z_0|$ to $F_0$.

To write down the general solution of (1.51) we first of all observe that it is possible to express $z_0$ as a product of the amplitude $|z_0|$ and a phase factor $e^{i\psi_0}$, i.e., $z_0 = |z_0|e^{i\psi_0}$, where $\psi_0$ can be put as

$$\psi_0 = tan^{-1}\frac{\lambda\omega\omega_0}{(\omega^2 - \omega_0^2)} \tag{1.58}$$

Implicit in the above is the feature that $\psi_0 \to \pm 0$ as $\omega$ approaches $\pm\infty$ while $\psi_0 \to \mp 0$ as $\omega$ approaches 0 from the positive or the negative side.

In this way, we are led to the following expression for $z(t)$:

$$z(t) = \frac{F_o}{m[(-\omega^2 + \omega_0^2)^2 + \lambda^2\omega^2\omega_0^2]^{\frac{1}{2}}} e^{i(\omega t + \psi_0)} \tag{1.59}$$

The general solution of the linear differential equation (1.51) is the superposition of the complementary function noted earlier in (1.45) for the damped oscillator problem and the one obtained by projecting out the real part from (1.59). Thus we arrive at

$$x(t) = Ae^{\frac{-\lambda\omega_0 t}{2}} \cos[\omega_0\sqrt{1 - \frac{\lambda^2}{4}}t + \phi] + \frac{F_o}{m[(-\omega^2 + \omega_0^2)^2 + \lambda^2\omega^2\omega_0^2]^{\frac{1}{2}}} \cos(\omega t + \psi_0) \tag{1.60}$$

A look at the above complete solution reveals that the first term, which contains the initial conditions, but being appended with a damping exponential term eventually dies out at large times and is therefore *transient* in character. Therefore, the motion is ultimately governed by the *steady* solution which being a particular solution of the differential equation is independent of the initial conditions:

$$x_{steady} = \frac{F_o}{m[(-\omega^2 + \omega_0^2)^2 + \lambda^2\omega^2\omega_0^2]^{\frac{1}{2}}} \cos(\omega t + \psi_0) \tag{1.61}$$

The amplitude of the steady state solution is given by $|z_0|$ and has a maximum value when the quantity $[(-\omega^2 + \omega_0^2)^2 + \lambda^2\omega^2\omega_0^2]$ reaches a minimum. The required condition for this to happen amounts to $\omega^2 = {\omega_R}^2 = \frac{1}{2}(2 - \lambda^2)\omega_0^2$ which is positive for $\lambda < 2$. Near $\omega = \omega_R$ results in a large response, in other words, very large oscillations and the behavior of resonance is noticed.

### (b) Forced harmonic oscillator with nonlinear damping

Finally, we consider the problem of a periodically forced oscillator subjected to a nonlinear damping according to the equation[1]

$$\ddot{x} + \omega_0^2 x + \lambda\omega_0|\dot{x}|\dot{x} = \frac{F_o}{m}\cos(\omega t) \tag{1.62}$$

We proceed in a somewhat different way than pursued in the previous case. First of all let us assume a solution like

$$x(t) = A\cos(\omega_0 t + \phi) \tag{1.63}$$

[1]A. Li, L. Ma, D. Keene, J. Klingel, M. Payne and X-J. Wang, Forced oscillations with linear and nonlinear damping, *Am. J. Phys.* 84(2016)32.

where $A$ is the amplitude. Substituting it in the above equation yields the expression

$$A(\omega_0^2 - \omega^2)\cos(\omega t + \phi) - \lambda\omega_0 A^2\omega^2|\sin(\omega t + \phi)|\sin(\omega t + \phi) \\ = \frac{F_o}{m}[\cos(\omega t + \phi)\cos\phi + \sin(\omega t + \phi)\sin\phi] \quad (1.64)$$

We now exploit the above relation by multiplying both sides by $\sin(\omega t + \phi)$ and integrate from 0 to $2\pi$ i.e., over a complete period. In this way we arrive at the relation

$$\frac{8\lambda\omega_0}{3\pi}A^2\omega^2 = -\frac{F_o}{m}\sin\phi \quad (1.65)$$

where we have used the definition $|\sin\theta| = \pm\sin\theta$ along with the trigonometric identity $4\sin^3\theta = 3\sin\theta - \sin 3\theta$. In the above equation the $+$ and $-$ signs correspond, respectively, to the intervals $0 < \theta < \pi$ and $\pi < \theta < 2\pi$.

On the other hand, multiplication of both sides of (1.64) by $\cos(\omega t + \phi)$ gives on integration from 0 to $2\pi$

$$A(\omega_0^2 - \omega^2) = \frac{F_o}{m}\cos\phi \quad (1.66)$$

The above two relations put us in a position to determine the amplitude $A$ and the phase angle $\phi$ in the forms

$$A = \frac{3\pi}{8\lambda\omega_0}\sqrt{-\frac{\Delta}{2} \pm \sqrt{\frac{\Delta^2}{4} + \frac{64\lambda^2\omega_0^2}{9\pi^2\omega^4}(\frac{F_o}{m})^2}}, \quad \tan\phi = \frac{8\lambda\omega_0 A}{3\pi\sqrt{\Delta}} \quad (1.67)$$

where $\Delta$ is given by $\Delta = [1 - (\frac{\omega_0}{\omega})^2]^2$.

---

## 1.7 The simple pendulum problem

As shown in Figure 1.3, in the simple pendulum case, a bob $P$ of mass $m$ is hung from a fixed point $O$ by an inextensible rod of length $l$ and is allowed to swing freely in a vertical plane only. Air resistance is neglected because it can be taken to be too small to influence the motion. Of the two forces at work, the tension is directed toward $O$ while the force of gravity always acts in the downward direction. The former does no work because it is operative in a direction perpendicular to the direction of motion of the bob.

The force driving the bob is given by $-mg\sin\theta$ where the negative sign is appended because $g$ acts downward. To write out the equation of motion, we note that for the displacement $y$ of the bob from the equilibrium state, we need to equate, by Newton's second law of motion, the driving force with

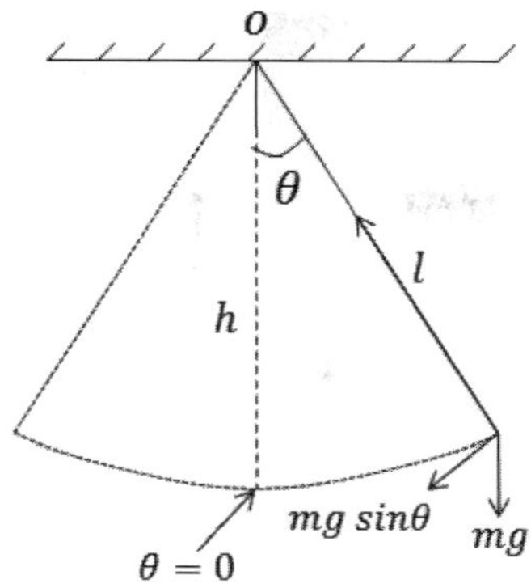


**FIGURE 1.3**: A simple pendulum swinging in the vertical plane.

the mass times acceleration $\ddot{y}$ of the displacement. This gives the relation $m\ddot{y} = -mg\sin\theta$. Now since $\theta$ is the angular displacement of the bob with the vertical axis, as it sweeps out a circular arc while it swings from one position to another, we have the relationship $y = l\theta$ from the arc length formula. Hence one deduces the equation of motion

$$\ddot{\theta} + \frac{g}{l}\sin\theta = 0. \tag{1.68}$$

A linearlized version of the above equation, when $\sin\theta$ is approximated by $\theta$, is the form of the simple harmonic oscillator whose general solution is sinusoidal in nature as noted in (1.33). Its period of oscillation is

$$\mathrm{T} = 2\pi\sqrt{\frac{l}{g}} \tag{1.69}$$

Since $T$ here is independent of the amplitude, the isochronicity of the simple pendulum is indicated in a first approximation.

Comparing (1.69) with the period of oscillation of the SHM as furnished by (1.35) we notice the correspondence

$$\sqrt{\frac{k}{m}} \longrightarrow \sqrt{\frac{g}{l}} \tag{1.70}$$

Numerically for the standard value of $g = 9.8062m/s^2$ and length of 1 meter, (1.69) yields a period of approximately 2 seconds. The general solution of (1.68) is governed by the Jacobi elliptic function to which we shall return later.

The potential energy of the simple pendulum reads

$$V = -\int_0^h (-mg)dx = mgh \tag{1.71}$$

where, as shown in Figure 1.3, $h$ corresponds to the height or the distance from the point of suspension to the resting or equilibrium position of the bob. Since $h = l - l\cos\theta$, V can be expressed as

$$V(\theta) = mgl(1 - \cos\theta) \tag{1.72}$$

where $\theta$ is the angle between $OP$ and the vertical line $OC$.

The total energy is given by

$$E = \frac{1}{2}ml^2(\frac{d\theta}{dt})^2 + mgl(1 - \cos\theta) \tag{1.73}$$

where the first term in the right-hand side represents the kinetic energy term. This is in conformity with the energy relation (1.39) of the harmonic oscillator taking into account the replacement (1.70), the connection of the angular displacement $\theta$ with $y$ namely, $y = l\theta$ and approximating $\cos\theta \approx 1 - \frac{\theta^2}{2}$.

---

## 1.8 Conservation principles

### 1.8.1 Conservation of linear momentum

Let a system of $N$ particles be acted upon by a set of external forces. We also take in the possibility that the particles may interact internally with each other because of the mutual forces operating between themselves. Ignoring the self forces the latter actions may be represented by $\vec{F_{ij}}$ where $\vec{F_{ii}} = 0, \quad i,j = 1,2,...N$. In this way we can project out the equation of motion of the $i^{th}$ particle in the manner

$$\frac{d\vec{p_i}}{dt} = m_i\ddot{\vec{r_i}} = \vec{F_i}^E + \sum_{i\neq j}\vec{F_{ij}}, \quad i,j = 1,2,...N \tag{1.74}$$

where $\vec{p_i}$ is the linear momentum of the $i^{th}$ particle, $\vec{F_i}^E$ is the total external force on the $i^{th}$ particle and the terms $\sum_{i\neq j}\vec{F^{ij}}$ denote the remaining ones due to the internal interactions: $j \neq i$ implying that self-interactions are being discarded.

Summing (1.74) over all the particles we get

$$\sum_i \frac{d\vec{p_i}}{dt} = \sum_i m_i\ddot{\vec{r_i}} = \sum_i \vec{F_i}^E + \sum_i\sum_{i\neq j}\vec{F_{ij}}, \quad i,j = 1,2,...N \tag{1.75}$$

Using Newton's third law, i.e., mutually interacting particles exert equal and opposite force forces upon themselves implying conditions like

$$\vec{F_{ij}} = -\vec{F_{ji}} \quad i,j = 1,2,...N \tag{1.76}$$

it follows that the double-sum in the right-hand side of (1.75) vanishes. We are therefore left with an equation like

$$\frac{d\vec{P}}{dt} = \sum_i m_i \ddot{\vec{r_i}} = \sum_i \vec{F_i}^E = \vec{F}, \quad i = 1, 2, ..., N \tag{1.77}$$

where $\vec{P}$ is the total linear momentum of the particles, pointing to the equality of the rate of change of the total linear momentum as equal to the total external force $\vec{F}$. If further we ignore the collective effect of the total external forces, then the last equation integrates to the result for the constancy of the linear momentum for a closed system: $\sum_i m_i \vec{v_i}$ = a constant vector, where $\vec{v_i} = \frac{d\vec{r_i}}{dt}$ is the $i^{th}$ velocity.

### 1.8.2 Conservation of angular momentum

Let $\vec{\Omega}$ be the resultant angular momentum of a system of particles

$$\vec{\Omega} = \sum_i \vec{r_i} \times \frac{d(m_i \vec{r_i})}{dt} \tag{1.78}$$

where the angular momentum of each particle has been summed over. We now compute the total derivative of $\vec{\Omega}$ to write down

$$\frac{d\vec{\Omega}}{dt} = \sum_i \vec{v_i} \times m_i \vec{v_i} + \sum_i \vec{r_i} \times m_i \ddot{\vec{r_i}} = \sum_i \vec{r_i} \times \vec{F_i^E} \tag{1.79}$$

where the first term in the right-hand side of the first equality vanishes because of the cross-product between similar vectors. For the second term which receives contributions from both the parts $\vec{F_i^E}$ and $\sum_{i \neq j} \vec{F_{ij}}$, $i, j = 1, 2, ...N$, yields, when summed over, only the quantity $\sum_i \vec{r_i} \times \vec{F_i^E}$ because other terms being of the type $\vec{r_1} \times \vec{F_{12}} + \vec{r_2} \times \vec{F_{21}}$ reduce to just $(\vec{r_1} - \vec{r_2}) \times \vec{F_{12}}$. But, by Newton's third law of motion, the sum of such vector products has to vanish because the force $\vec{F_{12}}$ acts along the line joining the particles 1 and 2.

The vector quantity in the right-hand side of (1.79) expressed in terms of the external force is called the vector moment or torque. It stands for the time rate of change of the angular momentum about a fixed point expressed in terms of the torques of all the external forces acting on the system. For a truly closed system the external forces have no role upon the system and one is led to the result that $\frac{d\vec{\Omega}}{dt}$ vanishes. This implies that the angular momentum is conserved.

A couple of remarks are in order. First, use of a different vector $\vec{r_1}$ which is in the same line of action of $\vec{F_i}$ will not change the torque because of the null condition $(\vec{r_1} - \vec{r_i}) \times \vec{F_{1i}} = 0$. Second, if there are two equal and opposite forces $\vec{F_i}$ and $-\vec{F_i}$ acting along the same line, then also we have trivially $\vec{r_1} \times \vec{F_i} + \vec{r_1} \times (-\vec{F_i}) = 0$.

### 1.8.3 Conservation of energy

The criterion (vi) for the classification of a conservative force easily follows from Newton's second law of motion. Indeed we find for a particle of mass $m$ moving with velocity $\vec{v}$

$$\int m\vec{v}.\frac{d\vec{v}}{dt}dt = \int \vec{F}.\vec{v}dt = \int \vec{F}.d\vec{r} = -\int (\nabla V).d\vec{r} = -\int dV \qquad (1.80)$$

where the force $\vec{F} = -\nabla V$ in terms of the potential energy function $V$. The above equation then points to the constancy of energy:

$$\frac{1}{2}m|\vec{v}|^2 + V = \text{constant} = E \qquad (1.81)$$

where the constant is the total energy which is the sum of kinetic and potential energies. Like we saw in the case of the harmonic oscillator in (1.39), the energy equation (1.81) for a general class of potential $V$ gives a useful insight on the restriction of a particle motion.

In one dimension, for a general $V(x)$, we can reexpress (1.81) in the form

$$p = m\dot{x} = \sqrt{2m(E - V)} \qquad (1.82)$$

It gives the integral

$$t = \sqrt{\frac{m}{2}} \int \frac{dx}{\sqrt{E - V(x)}} + \text{constant} \qquad (1.83)$$

which provides $t$ as a function of $x$.

As an application of the above equation let us consider the simple case of a constant force acting on a particle under a potential given by $V = -cx$ where the constant $c > 0$. We find from (1.83)

$$\text{t} = \sqrt{\frac{m}{2}} \int_a^x \frac{dx}{\sqrt{E + cx}} = \frac{1}{c}\sqrt{2m}(\sqrt{E + cx} - \sqrt{E + ca}) \qquad (1.84)$$

where $a$ is the position of the particle at $t = 0$. With the help of the energy equation $E = \frac{1}{2}mv_0^2 - ca$, where $v_0$ is the initial velocity, we straightforwardly obtain the usual distance-time relation $x(t) = a + v_0 t + \frac{1}{2m}ct^2$.

Since the kinetic energy of a moving particle is always positive, have the constraint $V(x) < E$ for the motion be finite. Here the particle is confined (i.e., trapped) in the region for which such an inequality holds. However, should $V(x)$ equal $E$, the particle comes to rest and for subsequent dynamics the direction of motion is reversed and the process is continued leading to periodic oscillations. Naturally solving for the equation $V(x) = E$ yields the turning points and the latter effectively prescribe the limits of the finite motion. Beyond $V(x) = E$ the motion leads to untrapped trajectories. Trapped and untrapped trajectories are separated by curves called *separatrices*.

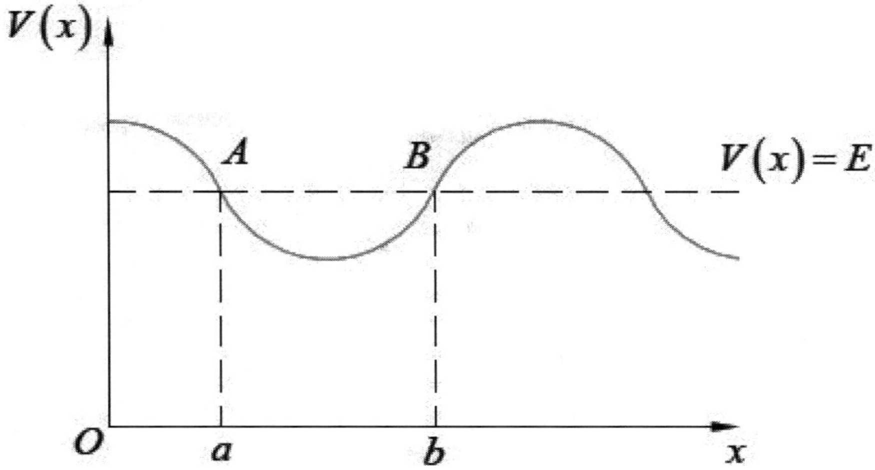


**FIGURE 1.4**: Restriction of the particle motion in the region $AB$ under the influence of the potential $V(x)$. The dotted line corresponds to $V(x) = E$.

In Figure 1.4 we have graphically illustrated the allowed region along with the turning points $A$ and $B$. It is easy to realize that for the motion bounded in the region $AB$, the particle moves back and forth between the turning points $a$ and $b$ and the motion is oscillatory in character. The period T is given by the time taken by the particle to travel from $A$ to $B$ along the potential curve and back. By symmetry this is twice the time from $a$ to $b$ and hence from (1.83) we obtain

$$\mathrm{T} = \sqrt{2m} \int_a^b \frac{dx}{\sqrt{E - V(x)}} \tag{1.85}$$

where the ranges of the integral correspond to the values of the roots of the equation $V(x) = E$ assuming $E$ to be known.

**Example 1.4**

For a particle under the influence of a quartic potential $V(x) = x^4$, the period of oscillation is

$$\mathrm{T} = 2\sqrt{2m} \int_0^{E^{\frac{1}{4}}} \frac{dx}{\sqrt{E - x^4}} \tag{1.86}$$

Substituting by turn $x = E^{\frac{1}{4}} z$ and $t = z^4$, it is straightforward to reduce the above integral to a tractable form

$$T = \sqrt{\frac{m}{2E}} E^{\frac{1}{4}} \int_0^1 t^{-\frac{3}{4}} (1-t)^{-\frac{1}{2}} dt \tag{1.87}$$

The integral in the right-hand side is in the form of the beta function integral $B(x, y) = \int_0^1 t^{x-1}(1-t)^{y-1} dt$ with $x = \frac{1}{4}$ and $y = \frac{1}{2}$. But since the beta function is expressible in terms of the $\Gamma$-function through $B(x, y) = \frac{\Gamma(x)\Gamma(y)}{\Gamma(x+y)}$, the time period for the quartic oscillator is given by

$$T = \frac{1}{2}\sqrt{\frac{2\pi m}{E}} E^{\frac{1}{4}} \frac{\Gamma(\frac{1}{4})}{\Gamma(\frac{3}{4})} \tag{1.88}$$

Consider again the problem of the simple pendulum. Here the total mechanical energy is conserved because the only external force, the tension T, does no work. Note that gravity is considered an internal force. More specifically, the conservation can be seen by calculating the time derivative of $E$ from (1.73) which gives

$$\frac{dE}{dt} = ml^2 \dot{\theta}(\ddot{\theta} + \frac{g}{l} \sin\theta) \tag{1.89}$$

The right-hand side vanishes by using the equation of motion of the pendulum implying that $E$ is constant with time.

Let us use the trigonometric identity $1 - \cos\theta = 2\sin^2(\frac{\theta}{2})$ to write the energy equation as

$$E(\theta) = \frac{1}{2} ml^2 (\frac{d\theta}{dt})^2 + 2mgl \sin^2(\frac{\theta}{2}) \tag{1.90}$$

As the pendulum oscillates in the vertical plane let $\theta_0$ be the highest point reached in the motion. Then $\dot{\theta}_0 = 0$ and we get from above

$$E(\theta_0) = 2mgl \sin^2(\frac{\theta_0}{2}) \tag{1.91}$$

Thus since the energy is conserved we can equate $E(\theta)$ and $E(\theta_0)$ to be led to the constraint

$$(\frac{d\theta}{dt})^2 = \frac{4g}{l}(\sin^2\frac{\theta_0}{2} - \sin^2\frac{\theta}{2}) \tag{1.92}$$

For small values of $\theta_0$, (1.92) can be approximated to

$$(\frac{d\theta}{dt})^2 \simeq \frac{4g}{l}(\frac{\theta_0^2}{4} - \frac{\theta^2}{4}) \tag{1.93}$$

which points to the equation of a circle

$$\theta^2 + (\frac{\dot{\theta}}{\sqrt{\frac{g}{l}}})^2 \simeq {\theta_0}^2 \tag{1.94}$$

in the plane of $\theta$ and $\frac{\dot{\theta}}{\sqrt{\frac{g}{l}}}$.

To obtain an estimate of the time period that follows from (1.93) let us first integrate it over one-fourth of the time period as the pendulum swings from the highest point $\theta_0$ to the zero value of $\theta$ at the vertical position

$$\int_0^{T/4} dt = \sqrt{\frac{l}{g}} \int_{\theta_0}^{0} \frac{d\theta}{\sqrt{\theta_0^2 - \theta^2}} \tag{1.95}$$

The integral being of the form $\sqrt{\frac{l}{g}} \sin^{-1}(\frac{\theta}{\theta_0})$ gives $\frac{\pi}{2}$ when evaluated over the end-points. We are therefore led to the same result of $T$ as obtained in (1.69).

In the next approximation it can be shown that

$$\mathrm{T} = \mathrm{T}_0(1 + \frac{1}{4}\sin^2\frac{\theta_0}{2}) \tag{1.96}$$

where $T_0$ is the value given by (1.69) (see Exercise 10).

The exact solution of (1.92) can be given by Jacobi elliptic functions. Toward this end let us introduce two variables $\eta$ and $\nu$ defined by

$$\eta = \sin(\frac{\theta}{2}), \quad \nu = \sin^2(\frac{\theta_0}{2}) \quad \Rightarrow \eta(0) = \sqrt{\nu} \tag{1.97}$$

Since

$$\frac{d\eta}{dt} = \frac{1}{2}\cos\frac{\theta}{2}\frac{d\theta}{dt}, \quad (\frac{d\eta}{dt})^2 = \frac{1}{4}(1-\eta^2)(\frac{d\theta}{dt})^2 \tag{1.98}$$

(1.92) is readily converted to the form in terms of the variable $\eta$

$$(\frac{d\eta}{dt})^2 = \nu\frac{g}{l}(1-\eta^2)(1-\frac{\eta^2}{\nu}) \tag{1.99}$$

Defining $\tau = \frac{g}{l}t$ and $\xi = \frac{\eta}{\sqrt{\nu}}$ one ends up with the differential equation

$$(\xi')^2 = (1-\xi^2)(1-\nu\xi^2), \quad 0 < \nu < 1 \tag{1.100}$$

where a prime denotes differentiation with respect to $\tau$. The general solution of (1.100) is one of Jacobi elliptic functions, namely, $\mathrm{sn}(\xi;\nu)$ with $\xi(0) = 1$ and $\xi'(0) = 0$ and $\nu$ is the modulus ranging between 0 and 1. For the end values of $\nu$, the Jacobi elliptic function reduces to $\sin\theta$ for $\nu = 0$ and $\tanh\theta$ for $\nu = 1$.

## 1.9 Perturbative analysis and the quartic oscillator

The method of approximations in connection with the traditional perturbation approach has a long history dating back to the times of Poincaré in the early part of the twentieth century. However, in the approximations of higher orders (i.e., truncating the series solution by a finite number of terms) one runs into difficulties because of the appearance of the secular or divergent terms.

For instance, consider the simple example of the differential equation (1.43) of the damped oscillator. The exact solution for the underdamping case governed by $\lambda < 2$ is furnished in (1.45) where the integration constants are determined from the initial conditions. If we look at $\lambda$ as a small parameter that could be treated perturbatively, then $x(t)$ can be expanded in a series form

$$x(t) = x_0(t) + \lambda x_1(t) + \lambda^2 x_2(t) + ... \tag{1.101}$$

For the zeroth-order and first-order of $\lambda$ we obtain the following respective equations:

$$\lambda^0(\text{zeroth-order}) \; : \; \ddot{x}_0 + \omega_0^2 x_0 = 0 \Rightarrow x_0 = A\cos(\omega_0 t + \phi) \tag{1.102}$$

$$\begin{aligned} \lambda^1(\text{first-order}) \; : \; & \ddot{x}_1 + \omega_0^2 x_1 = -\omega_0 \dot{x}_0 = A\omega_0^2 \sin(\omega_0 t + \phi) \\ & \Rightarrow x_1 = -\frac{A\omega t}{2}\cos(\omega t + \phi) \end{aligned} \tag{1.103}$$

Thus to $O(\lambda)$ we have the following approximate expression for $x(t)$

$$x = x_0 + \lambda x_1 = A\left(1 - \frac{\lambda\omega t}{2}\right)\cos(\omega t + \phi) \tag{1.104}$$

When compared with the exact solution (1.45), a disturbing feature of (1.104) is the presence of the term $\frac{A\lambda\omega t}{2}\cos(\omega t + \phi)$ which, due to the attachment of the factor $t$ in the coefficient, behaves as a secular or diverging term. It increases linearly with time. It is evident that such a perturbative treatment will inevitably be confronted by the appearance of such terms that blow up asymptotically. One way to avoid the secular term is to employ a power series to represent the nonlinear frequency as a function of $\lambda$ and then renormalize the coefficients for its removal. Such a procedure of renormalization is known as the Lindstedt–Poincaré method.

Let us apply such an approach for the quartic oscillator case. We note that the SHM deals with the case of an idealized spring when the system is guided by a linear force term. When the system suffers large amplitude oscillations there is indeed a deviation from such a situation. In effect this amounts to considering the higher effects in the expansion of the potential energy function given by (1.40) of which the quadratic term corresponds to the linear restoring force. Requiring the potential to be still symmetric under space inversion $x \to -x$, the quartic term in the potential is the next choice.

It corresponds to a cubic force that naturally makes the underlying equation of motion nonlinear:

$$\ddot{x} + \omega_0^2 x = -\gamma x^3 \tag{1.105}$$

(1.105) is called the equation of the *Duffing oscillator*. In the following we assume $\gamma > 0$.

To proceed further we adopt a change in variable:

$$\tau = \varpi t + \theta \tag{1.106}$$

where $\theta$ is a constant. It transforms according to

$$x(t) \to z(\tau) \equiv z(\varpi t + \theta) \tag{1.107}$$

As a result (1.105) takes the form

$$\varpi^2 \ddot{z} + \omega_0^2 z = -\gamma z^3, \quad \ddot{z} = \frac{d^2 z}{d\tau^2} \tag{1.108}$$

Writing down expansions of both $z(\tau)$ and the nonlinear frequency $\varpi$ in the form of a power series we obtain the first order in $\gamma$

$$z(\tau) = z_0(\tau) + \gamma z_1(\tau) + O(\gamma^2) \tag{1.109}$$

$$\varpi = \varpi_0 + \gamma \varpi_1 + O(\gamma^2) \tag{1.110}$$

Setting, as is customary in the power series approach, $\varpi_0 = \omega_0$ so that $\varpi^2 = \omega_0^2 + 2\gamma\omega_0\varpi_1 + O(\gamma^2)$, we get from (1.108) the following sequence of equations:

$$\gamma^0(\text{zeroth-order}) : {\omega_0}^2(\ddot{z}_0 + z_0) = 0 \tag{1.111}$$

$$\gamma^1(\text{first order}) : {\omega_0}^2(\ddot{z}_1 + z_1) = -z_0^3 - 2\omega_0\varpi_1\ddot{z}_0 \tag{1.112}$$

etc.

The zeroth-order resembles a SHM-like equation. The solution of (1.111) is of the form

$$z_0 = a\cos\tau = a\cos(\varpi t + \theta) \tag{1.113}$$

where $a$ is the amplitude. When the above solution is substituted in (1.112), it gives

$$\ddot{z}_1 + z_1 = -\frac{\Delta}{4{\omega_0}^2}\cos\tau - \frac{a^3}{4{\omega_0}^2} cos 3\tau \tag{1.114}$$

where the quantity $\Delta$ stands for $\Delta \equiv 3a^3 - 8a\omega_0\varpi_1 a \neq 0$. A particular solution of the above equation reads

$$z_1 = -\frac{\Delta}{16{\omega_0}^2}\cos\tau + \frac{a^3}{32{\omega_0}^2}\cos 3\tau - \frac{\Delta}{8{\omega_0}^2}\tau\sin\tau \tag{1.115}$$

Taken with (1.113), the general perturbative solution $z(\tau)$ up to $O(\gamma)$ is given by

$$z(\tau) = a\cos\tau + \gamma\left[-\frac{\Delta}{16{\omega_0}^2}\cos\tau + \frac{a^3}{32{\omega_0}^2}\cos 3\tau\right] - \frac{\Delta\gamma}{8{\omega_0}^2}\tau\sin\tau \qquad (1.116)$$

where the last term is the growth term.

In the Lindstedt–Poincaré approach, $\Delta$ is set equal to zero and the elimination of the secular term fixes $\varpi_1 = \frac{3}{8\omega_0}a^2$. It implies the dependence of the frequency on the amplitude. For the second-order, it is easily checked that one encounters three types of divergent terms $\tau\sin\tau, \tau\sin 3\tau$ and $\tau^2\cos\tau$. The first two terms vanish for $\varpi_1 = \frac{3}{8\omega_0}a^2$ and the second one for $\varpi_2 = \frac{3}{256\omega_0}a^4$. Thus, the elimination of the secular terms is done in each step of the power series in a recursive manner and renormalized expressions derived. However, a shortcoming of the perturbative method, which was developed mainly for astronomical calculations, is the difficulty one faces with the convergence of the power series employed although it must be said that such a disadvantage is not serious for a physical problem. Actually, what happens is that higher $\varpi_i$'s produce higher powers in $a$, i.e., $\varpi_1 \sim a^2, \varpi_2 \sim a^4$ etc., which means that for $a > 1$ the series diverges and the expansion does not make sense. In other words, the domain $a > 1$ cannot be covered within the canonical approach.

## 1.10 Rewriting Newton's second law in terms of kinetic and potential energy in a conservative system

Newton's second law of motion gives the relationship between the force, mass and acceleration. However, to put the law in practice it is often helpful to translate it in terms of the kinetic and potential energies of the system. For a configuration of $N$ particles with masses $m_1$, $m_2$,...$m_N$ situated at the points whose Cartesian coordinates are given by the respective sets ($x_1$, $y_1$, $z_1$),($x_2$, $y_2$, $z_2$),...,($x_N$, $y_N$, $z_N$), the total kinetic energy has the explicit form

$$\begin{aligned} T &= \frac{1}{2}(m_1\dot{x_1}^2 + m_2\dot{x_2}^2 + ... + m_N\dot{x_N}^2) + \frac{1}{2}(m_1\dot{y_1}^2 + m_2\dot{y_2}^2 + ... + m_N\dot{y_N}^2) \\ &\quad + \frac{1}{2}(m_1\dot{z_1}^2 + m_2\dot{z_2}^2 + ... + m_N\dot{z_N}^2) \end{aligned} \qquad (1.117)$$

Expressed as a sum, $T$ reads

$$T = \frac{1}{2}\sum_{i=1}^{N}(m_i\dot{x_i}^2 + m_i\dot{y_i}^2 + m_i\dot{z_i}^2) \qquad (1.118)$$

On the other hand, in a conservative system, the force acting on the system being derivable from a potential function $V(x)$ as its negative gradient, we can write for each component

$$F_{x_1} = -\frac{\partial V}{\partial x_1}, \quad F_{x_2} = -\frac{\partial V}{\partial x_2}, \quad ..., \quad F_{x_N} = -\frac{\partial V}{\partial x_N} \tag{1.119}$$

$$F_{y_1} = -\frac{\partial V}{\partial y_1}, \quad F_{y_2} = -\frac{\partial V}{\partial y_2}, \quad ..., \quad F_{y_N} = -\frac{\partial V}{\partial y_N} \tag{1.120}$$

$$F_{z_1} = -\frac{\partial V}{\partial z_1}, \quad F_{z_2} = -\frac{\partial V}{\partial z_2}, \quad ..., \quad F_{z_N} = -\frac{\partial V}{\partial z_N} \tag{1.121}$$

In a short-hand notation these are simply

$$F_{x_i} = -\frac{\partial V}{\partial x_i}, \quad F_{y_i} = -\frac{\partial V}{\partial y_i}, \quad F_{z_i} = -\frac{\partial V}{\partial z_i}, \quad i = 1, 2, ..., N \tag{1.122}$$

Employing the Cartesian components of the force-potential relationship as in above, Newton's laws for the $x$, $y$ and $z$ components can be put out in the manner

$$m_i\ddot{x}_i = -\frac{\partial V}{\partial x_i}, \quad m_i\ddot{y}_i = -\frac{\partial V}{\partial y_i}, \quad m_i\ddot{z}_i = -\frac{\partial V}{\partial z_i} \quad i = 1, 2, ..., N \tag{1.123}$$

From (1.118), since

$$\frac{\partial T}{\partial \dot{x}_i} = m_i\dot{x}_i, \quad \frac{\partial T}{\partial \dot{y}_i} = m_i\dot{y}_i, \quad \frac{\partial T}{\partial \dot{z}_i} = m_i\dot{z}_i \tag{1.124}$$

we can equivalently express (1.123) in terms of a set of the following three equations:

$$\frac{d}{dt}(\frac{\partial T}{\partial \dot{x}_i}) = -\frac{\partial V}{\partial x_i}, \quad \frac{d}{dt}(\frac{\partial T}{\partial \dot{y}_i}) = -\frac{\partial V}{\partial y_i}, \quad \frac{d}{dt}(\frac{\partial T}{\partial \dot{z}_i}) = -\frac{\partial V}{\partial z_i} \quad i = 1, 2, ..., N \tag{1.125}$$

In this way we have been able to express the Cartesian forms of Newtonian equations of motion in terms of the energies of the system rather than the forces themselves. To ensure a more compact representation of the above equations it is instructive to introduce a Lagrangian function defined as the difference of the kinetic energy (which is a function of velocities) and potential energy (which is a function of position in a conservative system) i.e., $L = T - V$. $L$ is thus a mixed function of coordinates and velocities.

Through the use of $L$, Equation (1.126) acquires the following forms:

$$\frac{d}{dt}(\frac{\partial L}{\partial \dot{x}_i}) = \frac{\partial L}{\partial x_i}, \quad \frac{d}{dt}(\frac{\partial L}{\partial \dot{y}_i}) = \frac{\partial L}{\partial y_i}, \quad \frac{d}{dt}(\frac{\partial L}{\partial \dot{z}_i}) = \frac{\partial L}{\partial z_i} \quad i = 1, 2, ..., N \tag{1.126}$$

Note that like the Newtonian equations, Lagrangian formulation also yields a set of second-order differential equations. The concept of a Lagrangian function is fundamental and taken as a reference point in the development of an analytical treatment of classical mechanics. Lagrangian equations can be derived in a more general setting by making use of D'Alembert's principle and getting rid of the so-called unknown forces of constraints (like the tension force in the simple pendulum problem which we identify as a constraint force) by invoking the principle of virtual work. This is the main advantage of Lagrangian mechanics. Note that Newton's laws do not distinguish the constraint forces as such. The avoidance of constraint forces and use of generalized coordinates (that could have a wide range of options given the nature of the system we are dealing with) are the key features in the Lagrangian description of things. For instance, a generalized coordinate could refer to an angle or an arc length apart from various other possibilities (the Cartesian or polar or cylindrical is one of many such possibilities) of definition. This of course leads to the minimization of the number of coordinates depending on the number of degrees of freedom that a physical system avails of.

Before concluding this section, let us revisit the problem of a simple pendulum as a typical application of Lagrangian dynamics. From (1.73) we can identify the kinetic energy and potential energy as

$$T = \frac{1}{2}ml^2(\frac{d\theta}{dt})^2, \quad V = mgl(1 - \cos\theta) \tag{1.127}$$

Here $\theta$ acts as the generalized coordinate and the Lagrangian function according to the definition $L = T - V$ is

$$L = \frac{1}{2}ml^2(\frac{d\theta}{dt})^2 - mgl(1 - \cos\theta) \tag{1.128}$$

Calculating the partial derivatives, namely,

$$\frac{\partial L}{\partial \theta} = -mgl\sin\theta, \quad \frac{\partial L}{\partial \dot{\theta}} = ml^2\dot{\theta} \tag{1.129}$$

and substituting them in a similar equation like (1.126) but interpreted in terms of the generalized coordinate $\theta$, the familiar equation of motion (1.68) for the simple pendulum straightforwardly emerges.

A detailed discussion of the Lagrangian formulation of classical mechanics will be taken up in Chapter 3 of the book.

---

## 1.11 Summary

By enunciating Newton's three laws of motion we addressed various aspects of classical mechanics that are inherent in these laws such as the concept of inertial frame, Galilean invariance and the homogeneous and isotropic character of the Newtonian Universe. We discussed the process of seeking the solutions of a physical system highlighting different coordinate frames that could be suitably employed to describe it. We looked, in particular, at the particular class of force called the conservative force which can be represented by a potential energy function. Use of the latter facilitates a great deal of simplification and its role was illustrated by turning to several problems that are of great importance in classical mechanics: the simple harmonic motion, the damped oscillator along with its various manifestations and the pendulum problem. Subsequently, the formulation of conservation principles was considered for the specific cases of linear momentum, angular momentum and energy. A general form of the time period formula was derived in terms of Jacobi elliptic function. We also introduced a treatment of perturbative analysis and discussed Lindstedt–Poincaré's technique of avoiding the secular terms in approximating a periodic solution of a differential equation focusing on the specific case of the quartic oscillator. Finally, by the introduction of kinetic and potential energies, we showed how a Lagrangian function can be defined for a conservative system in a Cartesian frame. A detailed treatment of the Lagrangian dynamics will be taken up in Chapter 3.

# Exercises

1. For the Earth whose radius at the equator is $R = 6400$ kms and angular velocity rotating on its axis given by $\omega = 7.3 \times 10^{-5}$ $rad/s$, show that the centripetal acceleration is 0.034 $m/s^2$.

2. A bird takes a spiral path to its nest in a manner that the radial distance reduces at a constant rate $r = a - bt$, while the angular speed increases at a constant rate $\dot{\theta} = \lambda t$. Show that the speed as a function of time is $v = \sqrt{b^2 + (a - bt)^2 \lambda^2 t^2}$.

3. Generalize the result (1.15) to the case of $N$-interacting particles.

4. Show that the expressions of the kinetic energy in spherical polar and cylindrical coordinates are respectively given by

$$T = \frac{1}{2} m(\dot{r}^2 + r^2 \theta^2 + r^2 \sin^2 \theta \dot{\phi}^2) \tag{1.130}$$

$$T = \frac{1}{2} m(\dot{\zeta}^2 + \zeta^2 \Phi^2 + \dot{z}^2) \tag{1.131}$$

5. Show that a particular solution of Laplace's equation $\nabla^2 \psi = 0$ which is independent of $\phi$ has a general solution

$$\psi = \sum_{n=1}^{\infty} (A_n r^n + B_n r^{-n-1}) P_n(\cos \theta) \tag{1.132}$$

where $P_n(\cos \theta)$ are the usual Legendre polynomials. Discuss the case when $\phi \neq 0$.

6. Identify the components of the momentum and angular momentum which are conserved in the field of an infinite homogeneous plane, infinite

homogeneous cylinder and infinite homogeneous semiplane which is bordered by one of the axis.

7. For a system of $N$ particles having an angular velocity $\omega$, determine the kinetic energy in the form $T = \sum_{i=1}^{n} \frac{1}{2} m_i \, |\vec{\omega} \times \vec{r_i}|^2$.

8. Show that in the case of the undamped oscillator the approximate expression of the total energy shows that it decays at half the rate at which amplitude does.

9. Find the period of oscillation for the potentials (i) $V(x) = -V_0 \cosh^{-2}(kx)$, (ii) $V(x) = \lambda |x|^n$ and (iii) $V(x) = V_0 \tan^2(kx)$.

10. If $\alpha$ denotes the amplitude of oscillation, show for the pendulum problem that the first correction to the time period of oscillation is given by

$$\mathrm{T} = 2\pi \sqrt{\frac{l}{g}} (1 + \frac{\alpha^2}{16} + O(\alpha^4)) \tag{1.133}$$

[Hint: Set $\bar{\omega} = \sqrt{\frac{g}{l}}$ and use the expansion $\sin\theta = \theta - \frac{\theta^3}{6}$ to write (1.68) as $\frac{d^2\theta}{dt^2} + \bar{\omega}^2(\theta - \frac{\theta^2}{6}) \simeq 0$. Try a solution $\theta = \bar{\theta}\cos\omega t + \mu\bar{\theta}^3 \cos 3\omega t$ to deduce $\omega = \bar{\omega}\sqrt{1 - \frac{1}{8}\bar{\theta}^2}$ and $\mu = -\frac{\bar{\omega}^2}{192}$. Hence derive the amplitude to this order of approximation.]

11. A particle of mass $m$ moving in the x direction has its kinetic energy given by $T = \frac{m}{2}\dot{x}^2$ and potential energy $V(x)$. Show that if $x = 0$ is a position of equlibrium, then for small values of $x$, the particle oscillates harmonically about the equilibrium point with an angular frequency $\omega = \sqrt{\frac{V''(0)}{m}}$.

12. Show that the equation for the damped linear oscillator can be expressed in a coupled equation form

$$\dot{x} = y, \quad \dot{y} = -\eta x - \omega^2 y \tag{1.134}$$

Interpret the function $\phi(x, y) = \frac{1}{2}(my^2 + kx^2)$ where $\eta = \frac{k}{m}$.

13. A particle of mass $m$ moves along the positive $x$-axis. It is acted upon by a constant force directed toward the origin with magnitude $C$ and and inverse square law repulsive force with magnitude $\frac{\mu}{x^2}$. Find the potential energy function $U(x)$ and the equilibrium position of the particle.

14. Consider the quartic oscillator potential $V(x) = \frac{1}{2}m\omega^2 x^2 + \frac{\lambda}{4}x^4$. Show for small $\lambda$ the time period is given by $T = \frac{2\pi}{\omega}(1 - \frac{3a^2\lambda}{6\omega^2})$ where $a$ is the amplitude of oscillation.

15. Consider the Lagrangian in the Cartesian coordinates

$$L = \frac{1}{2}m(\dot{x}^2 + \dot{y}^2 + \dot{z}^2) - mgz \tag{1.135}$$

Write down the equations of motions and all the conserved quantities that you can identify.

# Chapter 2

# Central force problems



The historical basis for central force problems comes from Kepler's three laws of planetary motion. These also go by the names of law of equal areas, law of ellipse and law of harmonics. Kepler announced the first two laws in 1609 while the third was stated in 1619 to get an insight into the behavior of the then known planets orbiting around the Sun. The statements of the three Kepler's laws are

1. Equal areas are swept out in equal intervals of time by the line segment drawn from a planet to the Sun.

2. The locus of a planetary orbit is an ellipse with the Sun at one of the foci.

3. The square of the periodic times of the planets is proportional to the cubes of the major axes of their orbits.

But we owe to Newton for a fuller understanding of these laws as explained in his *Principia* in 1687. Taking them in order, the law of equal areas focuses on the speed of the planet: it is fastest when it is closest to the Sun while it is slowest when it is most distant from the Sun. The law of ellipse points to the elliptical path that a planet traces out while moving around the Sun with the Sun in one of its foci. The law of harmonics provides a comparison of the orbital period (T) and radius of the orbit ($R$) as compared to other planets in the family. Quite interestingly, the ratio $\frac{\mathrm{T}^2}{R^3}$ turns out to be of nearly the same value of unity for every planet.

In this chapter we will be interested in the problem of the central force in which the force function depends *only* on the distance $r$ from the origin thus embodying the character of spherical symmetry. In some of the well-known examples of the central force like Newton's inverse square law of gravitation and Coulomb's electrostatic force between two charges, the force function $F(r)$ is proportional to the inverse square of the distance and is negative. However, in the case of the spherical harmonic oscillator, $F(r)$ is linearly proportional to the distance and negative as well. By Bertrand's theorem[1] it can be established that these two force functions are the only possible forms of the central force fields having orbits that are all stable and closed. In the central force problem the underlying force acts in a direction that is toward or away from a fixed point called the "force center." As such, the torque $\vec{r}\times\vec{F}$ on the particle about the force center vanishes resulting in the constancy of the angular momentum $\vec{\Omega} = \vec{r}\times(m\vec{v}) = \vec{r}\times\vec{p}$.

---

## 2.1 Inertial and gravitational mass: Principle of equivalence

We already discussed, in the previous chapter, the concept of the *inertial mass.* Let us compare the second law motion with Newton's inverse square law of gravitation between a body of mass $m$ at the position $\vec{r}$ and $n$ different others specified by their masses $m_i$ occupying positions $r_i$, $i = 1, 2, ..., n$. The force of attraction is given by

$$\vec{F} = -\sum_i \frac{GMm_i(\vec{r} - \vec{r_i})}{|\vec{r} - \vec{r_i}|^3}, \quad i = 1, 2, ..., n \tag{2.1}$$

where G is the gravitational constant approximately equal to $6.673 \times 10^{-11} Nm^2kg^{-2}$ and measured in Newtons ($N$), kilograms and meters.

What happens if we follow the second law of motion of Newton's to identify the $m$ in (2.1) as an inertial mass and write $\vec{F}$ as simply $\vec{F} = m\vec{a}$ ? In such a situation $m$ cancels out and we are left with an expression for the acceleration of a particle in a gravitational field:

$$\vec{a} = -G\sum_i \frac{m_i(\vec{r} - \vec{r_i})}{|\vec{r} - \vec{r_i}|^3}, \quad i = 1, 2, ..., n \tag{2.2}$$

To dig into this issue a little further, let us consider specifically two particles $m_1$ and $m_2$ in the presence of the gravitational field of a certain particle of mass $m$. We assume that both $m_1$ and $m_2$ are positioned at the same distance $r$ from $m$. In such a case, these particles would experience a force due to

[1]The essence of Bertrand's theorem is that the gravitational or electrostatic potential and the radial oscillator potential are the only types belonging to central force problems with the feature that all bound orbits are also closed. For a fuller discussion see E.T. Whittaker, *A Treatise on the Analytical Dynamics of Particles and Rigid Bodies*, Cambridge University Press, London (1988).

the gravitational field of $m$. We distinguish the role of mass $m$, which *exerts* a force, as an active gravitational mass and label it as $m_a$ while the receiver masses $m_1$ and $m_2$, which *experience* the force, as passive gravitational masses and label them as $m_{1p}$ and $m_{2p}$ . From (2.1) the two forces are given by

$$F_1 = -G\frac{m_a m_{1p}}{r^2}, \quad F_2 = -G\frac{m_a m_{2p}}{r^2} \tag{2.3}$$

In the context of Newton's second law of motion defining an inertial ($I$) mass, given the above two forces, we could define the corresponding accelerations $a_1$ and $a_2$ as given by

$$a_1 = \frac{F_1}{M_{1I}}, \quad a_2 = \frac{F_2}{M_{2I}} \tag{2.4}$$

in terms of two inertial masses $M_{1I}$ and $M_{2I}$.

At this point let us invoke the idea of the equivalence principle (weak) which states that if objects are dropped from some height from say, the top of a building, then so long as their motion is not affected by air resistance or other disturbances, will arrive at about the same time on the ground i.e., they fall at the same rate with equal accelerations. This curious feature is known from Galileo's time. (Mathematically, corresponding to $\vec{g}$, i.e., the acceleration due to gravity, assumed the same for particles near the Earth's surface, the force experienced by a particle of $m$ is $\vec{F} = m\vec{g}$. It is with this force that a particle coming under the influence of Earth's gravitational field will be attracted toward the Earth).

We therefore put $a_1 = a_2$ and divide out the two relations (2.3) and (2.4). It is straightforward to see that the active gravitational mass cancels out and we get the ratio

$$\frac{m_{1p}}{M_{1I}} = \frac{m_{2p}}{M_{2I}} \tag{2.5}$$

pointing to the proportionality between the passive and inertial masses.

In point of fact, the active and passive gravitational masses are also proportional among themselves. This can be seen[2] using Newton's third law of motion. Consider two particles at points $P_1$ and $P_2$ at respective distances $\vec{r_1}$ and $\vec{r_2}$ from a fixed point $O$ (see Figure 2.1). Then the force $F_1$ that the particle at $P_1$ exerts on $P_2$ and the force $F_2$ that the particle at $P_2$ exerts on $P_1$ are, respectively,

$$F_1 = -Gm_{2p}m_{1a}\frac{(\vec{r_2} - \vec{r_1})}{|\vec{r_1} - \vec{r_2}|^3}, \tag{2.6}$$

$$F_2 = -Gm_{1a}m_{2p}\frac{(\vec{r_1} - \vec{r_2})}{|\vec{r_1} - \vec{r_2}|^3}, \tag{2.7}$$

[2]We follow M.V. Berry, *Principles of Cosmology and Gravitation*, Cambridge University Press, London (1989).

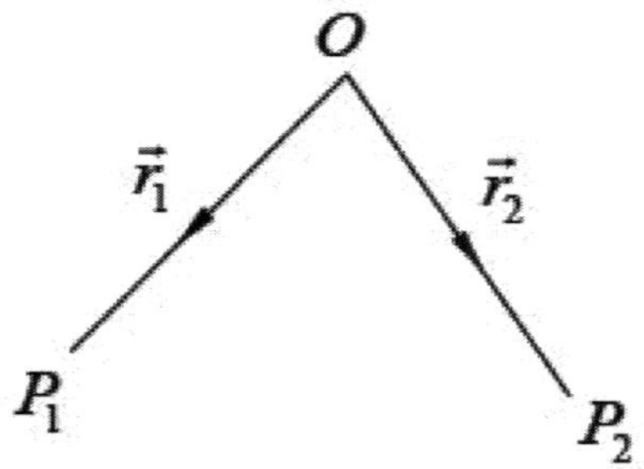


**FIGURE 2.1**: Two particles at points $P_1$ and $P_2$ with respect to the fixed point $O$.

The directions of $F_1$ and $F_2$ are opposite in accordance with Newton's third law but they have to be equal in magnitude as well. This requires the following ratio to hold if we divide the above two relations:

$$\frac{m_{1p}}{m_{2p}} = \frac{m_{1a}}{m_{2a}} \tag{2.8}$$

In other words, passive and active gravitational masses are proportional. Indeed there is no difference between the gravitational mass and inertial mass as experimental results have time and again shown. In his theory of General Relativity, Einstein also did not distinguish between gravitational and inertial mass. He asserted "*...we shall (therefore) assume the complete physical equivalence of a gravitational field and a corresponding acceleration of the reference system.*" He put out his idea of the principle of equivalence in the following way: "*We consider two systems $\Sigma_1$ and $\Sigma_2$ in motion. Let $\Sigma_1$ be accelerated in the direction of its $X$-axis, and let $\gamma$ be the (temporally constant) magnitude of that acceleration. $\Sigma_2$ shall be at rest, but it shall be located in a homogeneous gravitational field that imparts to all objects an acceleration $-\gamma$ in the direction of the $X$-axis. As far as we know, the physical laws with respect to $\Sigma_1$ do not differ from those with respect to $\Sigma_2$; this is based on the fact that all bodies are equally accelerated in the gravitational field.*"

---

## 2.2 Derivation of Kepler's three laws

We focus on the two-body system of the Sun and Earth whose respective masses are given by say, $M$ and $m$. An implication of the identification of $m$ as the inertial mass is that from (2.2) we can express $\vec{a}$, for such a system, as

$$\vec{a} = -\frac{GM}{r^3}\vec{r} \tag{2.9}$$

Clearly (2.9) is the outcome of the inverse square force law of gravitation.

Since $\frac{d(\vec{r}\times\frac{d\vec{r}}{dt})}{dt} = \vec{r}\times\vec{a}$ and $\vec{a}$ being given by above as directly proportional to $\vec{r}$ , we are led to the vanishing of the time derivative of the quantity $\vec{r}\times\frac{d\vec{r}}{dt}$ implying that the latter is a constant vector $\vec{C}$:

$$\vec{C} = \frac{\vec{l}_E}{m} \tag{2.10}$$

where $\vec{l}_E$ is the angular momentum of Earth about the Sun. Thus $\vec{l}_E$ is a conserved quantity. As such

$$\vec{r}\times\frac{d\vec{r}}{dt} = \vec{C} = \frac{\vec{l}_E}{m} \tag{2.11}$$

The sectorial area $A$ being given by $\frac{1}{2}|\vec{r}\times d\vec{r}|$, it follows from (2.10) and (2.11) that

$$2\frac{dA}{dt} = \vec{C} = |\frac{\vec{l}_E}{m}| \tag{2.12}$$

In other words, equal areas are swept out in equal intervals of time which is Kepler's first law. That the motion lies on a plane is because $\vec{r}$ remains perpendicular to $\vec{l}_E$:

$$\vec{r}\cdot[\vec{r}\times\frac{d\vec{r}}{dt}] = \vec{r}\cdot\frac{\vec{l}_E}{m} = 0. \tag{2.13}$$

Next from (2.9) and (2.11) we can write

$$\frac{d(\vec{C}\times\frac{d\vec{r}}{dt})}{dt} = \vec{C}\times\vec{a} = \frac{GM}{r^2}\vec{r}\times(\vec{r}\times\frac{d\vec{r}}{dt}) \tag{2.14}$$

Using the vector triple product formula

$$\vec{A}\times(\vec{B}\times\vec{C} = \vec{B}(\vec{A}\cdot\vec{C}) - \vec{C}(\vec{A}\cdot\vec{B}) \tag{2.15}$$

on integrating (2.14) we obtain

$$\vec{C}\times\frac{d\vec{r}}{dt} = -(\frac{GM}{r}\vec{r}+\vec{D}) \tag{2.16}$$

where $\vec{D}$ is some constant vector. Taking now a dot product with $\vec{r}$, (2.16) becomes

$$(\vec{r}\times\frac{d\hat{r}}{dt})\cdot\vec{C} = GMr + \vec{r}\cdot\vec{D} \tag{2.17}$$

Since $\vec{D}$ is arbitrary, we can choose its direction to coincide with the polar axis. Thus we can represent $\vec{r}\cdot\vec{D} = rD\cos\theta$. Using now (2.11) we easily derive the polar representation of a conic as given by

$$r = \frac{\varrho}{1+e\cos\theta} \tag{2.18}$$

where the parameters $\varrho$, $e$ are $\varrho = \frac{l_E{}^2}{GMm^2}$ and $e = \frac{D}{GM}$. For a planetary orbit, (2.18) is a closed curve such as an ellipse with the focus at the origin and $e$ its eccentricity. In this way we arrive at a mathematical form of Kepler's second law.

The points $\theta = 0$ and $\theta = \pi$ are the ends where (2.18) crosses the polar axis. Adding the corresponding values of $\theta$ we determine the length $2a$ of the major axis of the ellipse to be

$$2a = \varrho(\frac{1}{1+e} + \frac{1}{1-e}) = 2\frac{C^2}{GM(1-e^2)} \tag{2.19}$$

where $\varrho$ has been expressed in terms of $C$ by using (2.10). Now the area of an ellipse being $A = \pi ab$, where $b$ is related to the length $a$ of the semi-major axis by $b = a\sqrt{1-e^2}$ and, from (2.12), the period for one complete revolution to be given by $\mathrm{T} = \frac{2A}{C}$, we obtain, on substituting for $C$ from (2.19), the formula

$$\mathrm{T} = 2\pi\sqrt{\frac{a^3}{GM}} \tag{2.20}$$

The result (2.20) has the implication that $\frac{\mathrm{T}^2}{a^3}$ behaves as a constant given by $\frac{4\pi^2}{GM}$ which corresponds to Kepler's third law.

---

## 2.3 Properties and equations of orbits

For a general force law $\vec{F}$ which is central, with the particle moving in a plane, it acts in a direction which is toward or away from a fixed point called the *force center*. As such, the *torque* $\vec{r} \times \vec{P}$ on the particle about the force center vanishes resulting in the constancy of the angular momentum $\vec{l}$:

$$\vec{l} = \vec{r} \times (m\ddot{\vec{r}}) \tag{2.21}$$

The conservation of the angular momentum has the implication that the orbital plane, given by the equation $\vec{l} \cdot \vec{r} = 0$, is the one on which the motion is confined and is normal to the angular momentum vector $\vec{l}$. Of course to determine the motion completely we also need to define the initial position vector $\vec{r}(0)$ and the initial velocity vector $\vec{v}(0)$ on such a plane.

Let us assume $|\vec{l}| \neq 0$. If $(r, \theta)$ are the coordinates of the particle in the orbital plane with respect to a fixed origin, then Newton's equations of motion in plane polar coordinates read

$$m(\ddot{r} - r\dot{\theta}^2) = F(r), \quad m(r\ddot{\theta} + 2\dot{r}\dot{\theta}) = 0 \tag{2.22}$$

The second equation integrates to

$$\frac{1}{r}\frac{d}{dt}(mr^2\dot{\theta}) = 0 \tag{2.23}$$

pointing to the constancy of the angular momentum $l = m\omega r^2$ with respect to time where $\omega = \dot{\theta}$. In terms of $\omega$, the radial equation becomes $\ddot{r} = r\omega^2 + \frac{1}{m}F(r)$ and looks like a typical representative of a one-dimensional motion but restricted on the half line $0 < r < \infty$.

It is often instructive to employ a change of variable $r = \frac{1}{u}$ to effect the transformations

$$\dot{r} = -\frac{l}{m}\frac{du}{d\theta}, \qquad \ddot{r} = -\frac{l^2u^2}{m^2}\frac{d^2u}{d\theta^2} \tag{2.24}$$

which cast the radial equation in the form

$$\frac{d^2u}{d\theta^2} + u = -F(u^{-1})\frac{m}{l^2u^2} \tag{2.25}$$

(2.25) provides the differential equation of the orbit for a particle under the control of a central force of a general form $P(\frac{1}{u})$.

Sometimes it is useful to display the angular momentum variable explicitly. In the presence of a potential $V(r)$ given by $F(r) = -\frac{dV}{dr}$, the radial equation acquires the form

$$m\ddot{r} = -\frac{d}{dr}[\frac{l^2}{2mr^2} + V(r)] \longrightarrow \quad m\ddot{r} = -\frac{dU(l,r)}{dr} \tag{2.26}$$

and we see that the motion is influenced by an effective potential

$$U(l,r) = \frac{l^2}{2mr^2} + V(r) \tag{2.27}$$

The first term in the right-hand side of (2.27) is the effect of a centrifugal barrier whose magnitude is given by $\frac{l^2}{mr^3}$.

Integrating (2.26) with respect to time $t$, the total energy $E$ that is constant in time is given by

$$E = \frac{1}{2}m|\dot{\vec{r}}|^2 + \frac{l^2}{2mr^2} + V(r) = \frac{1}{2}m|\dot{\vec{r}}|^2 + U(l,r) \tag{2.28}$$

where $\frac{l^2}{2mr^2}$ is the centrifugal term $V_{cf}$. For the Newtonian inverse square law, $V(r) = -\frac{\gamma}{r}$ (which is negative for an attractive force, $\gamma > 0$), a graphical description of $U(l,r)$ and $V_{cf}$ is illustrated in Figure 2.2. We see that while $V_{cf}$ goes as $\frac{1}{r^2}$, $V(r)$ is always negative. As a result, $U(l,r)$ can descend to a minimum having a finite negative value.

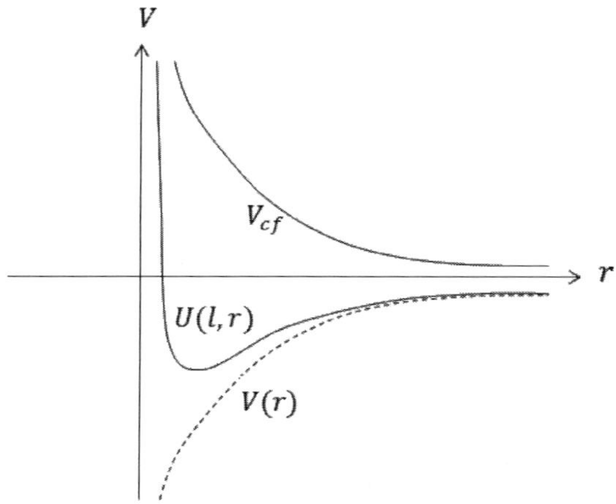


**FIGURE 2.2**: A graphical description of the function $U(l,r)$ against $V_{ef}$ and $V(r)$.

## 2.4 Integral representations

In terms of the angular momentum parameter $l$, the energy formula (2.28) can be easily represented in a derivative form

$$\frac{dr}{d\theta} = \pm \frac{mr^2}{l} \sqrt{\frac{2}{m}[E - U(l,r)]} \tag{2.29}$$

The corresponding representations for $u$ are

$$\frac{du}{d\theta} = \mp \frac{\sqrt{2m}}{l} \sqrt{E - V(\frac{1}{u}) - \frac{l^2 u^2}{2m}} \tag{2.30}$$

The choice of signs is guided by the initial conditions that are relevant to the system. Before we turn to the evaluation of the above integrals let us first consider a general class of power law potentials.

## 2.5 A general class of power law potentials

The entire class of power-law potentials can be described by the general form[3]

$$V(r) = -\frac{\gamma}{r^{\nu}} = -\frac{\gamma}{r^{2\lambda+2}}, \quad \nu \neq 0, \gamma \neq 0 \tag{2.31}$$

[3]J. Daboul and M. M. Nieto, Quantum bound states with zero binding energy, *Phys. Lett. A* 190 (1994) 357.

where $\lambda = \frac{1}{2}(\nu - 2)$. The above potential is attractive for $\nu > 0$ but repulsive for $\nu < 0$. Corresponding to (2.31), $U(l, r)$ is given by

$$U(l,r) = \frac{l^2}{2mr^2} - \frac{\gamma}{r^\nu} \tag{2.32}$$

$$\text{i.e.} \quad U(\rho) = \frac{l^2}{2ma^2}\left(\frac{1}{\rho^2} - \frac{1}{\rho^\nu}\right), \quad a = \left(\frac{2m\gamma}{l^2}\right)^{1/(\nu-2)} \tag{2.33}$$

where $\rho = \frac{r}{a}$ and $a$ is determined from the condition $U_p(l, a) = 0$.

In terms of $a$, $V(r)$ assumes the form

$$V(r) = -\frac{l^2 a^{\nu-2}}{2m}\frac{1}{r^\nu} \tag{2.34}$$

We speak of a bounded motion if $r = r_{\text{min}}$ and $r = r_{\text{max}}$ exist where $\dot{r} = 0$. From (2.28), since $E = \frac{1}{2}m\dot{r}^2 + U(l, r)$ for all $r$, we have the condition $U(l, r) \leq E$ for a physical bounded motion. In Figure 2.2, the curve $U(l, r)$ is seen to descend to a minimum with a finite negative value implying a range of bounded orbits.

Among the orbits the circular ones are, from the mathematical point of view, the simplest to pick. For a particle in circular motion of radius $r = \bar{r}$ with a constant acceleration $\frac{v^2}{\bar{r}}$ toward its center, $v$ being the magnitude of its velocity given by $v = \bar{r}\dot{\theta}$, it is clear that the following balancing equation holds:

$$F(\bar{r}) + m\bar{r}\dot{\theta}^2 = 0$$

$$\text{i.e.} \quad F(\bar{r}) + \frac{l^2}{m\bar{r}^3} = 0 \tag{2.35}$$

where $F(\bar{r})$ is the force at $r = \bar{r}$.

For the stability of the circular orbit we enquire for a minimum of the effective potential $U(l, r)$ and hence for the following two conditions to hold:

$$U'(l,r) \equiv \frac{dU(l,r)}{dr} = 0, \quad U''(l,r) \equiv \frac{d^2U(l,r)}{dr^2} > 0 \tag{2.36}$$

at $r = \bar{r}$. Note that (2.35) is also consistent with seeking a derivative of $U(l, r)$ with respect to $r$ and putting $r = \bar{r}$ remembering that the force $F(r)$ is given by the negative gradient of the potential.

Writing

$$m\ddot{r} = m\frac{d^2}{dt^2}(r - \bar{r}) = -U'(l,r) \tag{2.37}$$

where $U'(l, r) = \frac{dU(l,r)}{dr}$, let us expand the right-hand side as

$$\begin{aligned} U'(l,r) &\approx U'(l,\bar{r}) + (r - \bar{r})U''(l,\bar{r}) \\ &= (r - \bar{r})U''(l,\bar{r}) \end{aligned}$$

This gives

$$m\frac{d^2}{dt^2}(r-\bar{r}) \approx -(r-\bar{r})U''(l,\bar{r}) \tag{2.38}$$

which we recognize as the familiar equation of the undamped oscillator, implying stability of the circular orbit for $U''(l,\bar{r}) > 0$. Since

$$U'(l,\bar{r}) = -\frac{l^2}{m\bar{r}^3} + V'(\bar{r}), \quad U''(l,\bar{r}) = \frac{3l^2}{m\bar{r}^4} + V''(\bar{r}) \tag{2.39}$$

we then have a small amplitude oscillation with an angular frequency $\omega$ given by

$$\omega = \sqrt{\frac{U''(l,\bar{r})}{m}} = \sqrt{\frac{V''(\bar{r}) + 3l^2/m\bar{r}^4}{m}} \tag{2.40}$$

If we eliminate $l$ from the condition $U'(l,\bar{r}) = 0$ then we have another representation for $\omega$

$$\omega = \sqrt{\frac{1}{m}\left[V''(\bar{r}) + \frac{3V'(\bar{r})}{\bar{r}}\right]} \tag{2.41}$$

The period of the orbit is given by

$$T = \frac{2\pi}{\sqrt{\frac{1}{m}\left[V''(\bar{r}) + \frac{3V'(\bar{r})}{\bar{r}}\right]}} \tag{2.42}$$

As an application let us examine the stability condition for the circular orbits of the entire class of power law potentials as given by (2.31).

In such a context the force corresponding to $V(r)$ must be attractive, i.e.,

$$F = -\frac{dV}{dr} < 0.$$

and implies $-\nu\gamma r^{-\nu-1} < 0$. So we must have $\nu\gamma > 0$.

The condition of stability of a circular orbit as just deduced is $U''(l,\bar{r}) > 0$. From (2.40) and (2.41) it gives $V''(\bar{r}) + \frac{3V'(\bar{r})}{\bar{r}} > 0$. As a consequence we work out the condition

$$-\gamma\nu(\nu+1)\bar{r}^{-\nu-2} + \frac{3}{\bar{r}}(\gamma\nu)\bar{r}^{-\nu-1} > 0$$

$$\text{or,} \quad -\gamma\nu[(\nu+1) - 3] > 0$$

$$\text{or,} \quad -\gamma\nu(\nu-2) > 0$$

Since $\gamma\nu > 0$ we are led to the condition $\nu < 2$.

## 2.6 Mapping the general class of potentials: Orbit equation for the inverse square law problem

We start with the general case of the power law potentials as given by (2.31). We deduce[4] from (2.29) the corresponding integral form for $\theta(r)$:

$$\theta(r) = \int \frac{dr}{r^2\sqrt{\frac{2m}{l^2}(E + \gamma r^{-\nu}) - \frac{1}{r^2}}} \tag{2.43}$$

Making a change of variable by setting $u = r^{\frac{\nu}{\alpha}}$ in (2.31) where $\alpha$ is a real and arbitrary constant gives $dr = \frac{\alpha}{\nu}(u^{-1+\frac{\alpha}{\nu}})du$. As a result the integral is transformed to

$$\theta(u) = \frac{\alpha}{\nu}\int \frac{du}{u^2\sqrt{\frac{2m}{l^2}(Eu^{-2+2\frac{\alpha}{\nu}} + \gamma u^{-2-\alpha+2\frac{\alpha}{\nu}}) - \frac{1}{u^2}}} \tag{2.44}$$

in terms of the $u$-variable. Hence to put (2.44) into a correspondence with (2.30), we have to assume that the constants $\alpha$ and $\nu$ satisfy the constraint

$$2\frac{\alpha}{\nu} - \alpha = 2 \tag{2.45}$$

Note that it can also be re-expressed in an equivalent factorized form $(2 - \nu)(\alpha + 2) = 4$. As a result (2.44) assumes the representation

$$\theta(u) = \frac{\alpha}{\nu}\int \frac{du}{u^2\sqrt{\frac{2m}{l^2}(Eu^{\alpha} + \gamma) - \frac{1}{u^2}}} \tag{2.46}$$

The above form is concerned with the power law potential $u^{\alpha}$. Up to certain adjustments of the constants it compares interestingly with the representation (2.43) pointing to a kind of duality between the potentials $r^{-\nu}$ and $u^{\alpha}$.

## 2.7 Coulomb and isotropic oscillator potentials

The orbit equation for the Coulomb potential, $V(r) = -\frac{\gamma}{r}$, which corresponds to $\nu = 1$ in (2.31), is straightforwardly determined by integrating (2.26). We can equivalently integrate (2.25). It gives the following differential equation for $u$:

$$\frac{d^2u}{d\theta^2} + u = \frac{m\gamma}{l^2} \tag{2.47}$$

[4] A. K. Grant and J. L. Rosner, Classical orbits in power-law potentials *Am. J. Phys.* 62(1994)310.

Its general solution is given by

$$u = A\cos(\theta - \theta_0) + \frac{m\gamma}{l^2} \tag{2.48}$$

in terms of two constants $A$ and $\theta_0$. Without loss of generality we can set $\theta_0 = 0$ by assuming it to be measured from the pericentron (or pericenter). Pericentron is the term used for the shorter apsis while apocentron (or apocenter) stands for the longer apsis while dealing with the motion of orbits. Apsis is the point where $\frac{dr}{d\theta} = 0$.

Inverting $u$ in terms of $r$ yields an equation of a conic in the standard polar form

$$r = \frac{l^2}{m\gamma}\frac{1}{1 + e\cos\theta} \tag{2.49}$$

where the parameter $e$ stands for the eccentricity of the orbit. For the pericentron ($\theta = 0$) and apocentron ($\theta = \pi$) we have, respectively, the relations

$$r_{per} = \frac{l^2}{m\gamma}\frac{1}{1+e}, \quad r_{apo} = \frac{l^2}{m\gamma}\frac{1}{1-e} \tag{2.50}$$

Different cases of orbits arise depending on the various possibilities of $e$. Before we distinguish them it is worthwhile to have a look at the energy equation (2.28). For the Coulomb potential the expression for $E$ reads

$$E = \frac{1}{2}m|\dot{\vec{r}}|^2 + \frac{l^2}{2mr^2} - \frac{\gamma}{r} \tag{2.51}$$

and implies an effective potential

$$U(l, r) = \frac{l^2}{2mr^2} - \frac{\gamma}{r} \tag{2.52}$$

It gives for the rate of change of radial velocity

$$\frac{dr}{dt} = \pm\sqrt{\frac{2}{m}(E - U(l, r))} \tag{2.53}$$

We therefore conclude that the maximum radial velocity is attained when $U(l, r)$ is a minimum.

Writing $\frac{dr}{dt} = \frac{dr}{d\theta}\frac{d\theta}{dt}$, one can transform $E$ to a form

$$E = \frac{l^2}{2mr^2}[\frac{1}{r^2}(\frac{dr}{d\theta})^2 + 1 - 2\frac{m\gamma}{l^2}r] \tag{2.54}$$

Evaluating $\frac{dr}{d\theta}$ from (2.49) which gives

$$\frac{dr}{d\theta} = \frac{l^2}{m\gamma}\frac{e\sin\theta}{(1 + e\cos\theta)^2} \tag{2.55}$$

and using the explicit form for $r$, a little calculation yields a rather simple representation for $E$:

$$E = \frac{m\gamma^2}{2l^2}(e^2 - 1) \tag{2.56}$$

Notice no $\theta$-dependence in (2.56) which is as it should be because energy is conserved for the system under consideration. The interplay between the eccentricity $e$ and the energy is thus evident in determining the character of the conic. We summarize the different possibilities for the orbits:

$$\begin{aligned}
&\text{Ellipse (bounded or trapped orbits)} : e < 1 \Longrightarrow E < 0,\\
&\text{Circle (special case of an ellipse)} : e = 0 \Longrightarrow E < 0,\\
&\text{Hyperbola (unbounded orbits)} : e > 1 \Longrightarrow E > 0,\\
&\text{Parabola (open orbit)} : e = 1 \Longrightarrow E = 0.
\end{aligned}$$

Let us consider them in turn. To this end it is often useful[5] to transform (2.49) to Cartesian coordinates. With this aim in mind we put $x = r\cos\theta, y = \sin\theta$ which for the ellipse results in the standard canonical form encountered in coordinate geometry

$$\frac{(x + \frac{l^2}{m\gamma}\frac{e}{1-e^2})^2}{a^2} + \frac{y^2}{b^2} = 1 \tag{2.57}$$

where $a^2 = \frac{(\frac{l^2}{m\gamma})^2}{(1-e^2)^2}$ and $b^2 = \frac{(\frac{l^2}{m\gamma})^2}{(1-e^2)}$. Since for the ellipse $e < 1$, it follows that $b^2 < a^2$. Note that the geometric center of the closed orbit of the ellipse is not at the origin but shifted to the point $x = -\frac{l^2}{m\gamma}\frac{e}{1-e^2}$. We have here a periodic motion in which the radial distance moves between $r_{apo}$ and $r_{per}$. It can be shown that one of the foci is at the origin (see Exercise 3).

From (2.56) we can read off for the range of the energy

$$-\frac{m\gamma^2}{2l^2} < E < 0. \tag{2.58}$$

Note that the circle is an extreme case of the ellipse to which the latter degenerates to when $e = 0$. Here $E = -\frac{m\gamma^2}{2l^2}$. In the case of gravitation the circular orbit is due to a balance between the centrifugal force and the gravitational force. The radial velocity has a zero-value and there is only the transverse circular velocity having the value $\sqrt{\frac{\gamma}{a}}$, where $a$ is the semi-major axis of the ellipse. See Exercise 4.

Unlike an ellipse, the hyperbola describes the case of an open orbit. Its equation reads

$$\frac{(x - \frac{l^2}{m\gamma}\frac{e}{e^2-1})^2}{a^2} + \frac{y^2}{b^2} = 1 \tag{2.59}$$

[5] D. Tong, Lecture notes on dynamics, Cambridge lecture notes (unpublished).

where $a^2 = \frac{(\frac{l^2}{m\gamma})^2}{(e^2-1)^2}$ and $b^2 = \frac{(\frac{l^2}{m\gamma})^2}{(e^2-1)}$. The eccentricity value being greater than unity implies $E > 0$.

Finally for the parabola the Cartesian form of the equation is given by

$$y^2 = -2\frac{l^2}{m\gamma}x + (\frac{l^2}{m\gamma})^2 \tag{2.60}$$

With the eccentricity value of unity it corresponds to zero-energy orbits. A parabolic orbit is also open like the hyperbolic one.

We next address the three-dimensional oscillator having the same frequency in every direction. Such an oscillator is called the isotropic oscillator and is guided by the potential

$$V(r) = \frac{1}{2}m\omega^2 r^2 \tag{2.61}$$

It induces an effective potential

$$U(l,r) = \frac{1}{2}m\omega^2 r^2 + \frac{l^2}{2mr^2} \tag{2.62}$$

The minimum of $U(l,r)$ is estimated in the usual way. It turns out that the lowest point of $U(l,r)$ is reached at

$$r_{min} = \frac{|l|}{m\omega} \tag{2.63}$$

where $\frac{d^2U(l,r)}{dr^2} > 0$, $r_{min}$ is the point of minimum. The particle has a natural tendency to be stable at such a point. The energy at $r = r_{min}$ is $E_{min} = \frac{1}{2}\omega|l|$. All other values of energy are double-valued and greater than $E_{min}$. These values could be solved from (2.28) by putting $r^2 = t$ yielding the equation

$$m^2\omega^2 t^2 - 2mEt + l^2 = 0 \tag{2.64}$$

The two solutions of $t$ for such a quadratic equation are

$$t_+ = \sqrt{\frac{E}{m\omega^2}(1 - \sqrt{1 - \frac{E_{min}^2}{E^2}})}, \quad t_- = \sqrt{\frac{E}{m\omega^2}(1 + \sqrt{1 - \frac{E_{min}^2}{E^2}})} \tag{2.65}$$

To derive an orbit equation for the isotropic oscillator we can employ either (2.29) or (2.30). We find it preferable to use (2.30).

Note that for the isotropic oscillator potential we can express

$$E - V(\frac{1}{u}) - \frac{l^2u^2}{2m} = \frac{1}{u^2}[(m\frac{E^2}{2l^2} - \frac{1}{2}m\omega^2) - (\frac{lu^2}{\sqrt{2m}} - \frac{E}{l}\sqrt{\frac{m}{2}})^2] \tag{2.66}$$

Setting for the quantity $m\frac{E^2}{2l^2} - \frac{1}{2}m\omega^2 = \xi^2$ and making a change of variable $w = \frac{lu^2}{\sqrt{2m}}$ - $\frac{E}{l}\sqrt{\frac{m}{2}}$, the integral in (2.30) can be transformed to

$$\theta = -\frac{1}{2}\int \frac{1}{\sqrt{\xi^2 - w^2}}dw = \frac{1}{2}\arccos(\frac{w}{\xi}) \tag{2.67}$$

up to an additive constant. With $w = \xi \cos 2\theta$ we get for $r^2$ the expression

$$r^2 = \frac{\frac{l}{\sqrt{2m}}}{\frac{E}{l}\sqrt{\frac{m}{2}} + \sqrt{\frac{mE^2}{2l^2} - \frac{1}{2}m\omega^2 \cos 2\theta}} = \frac{1}{\frac{mE}{l^2}(1 + \sqrt{1 - \frac{l^2\omega^2}{E^2}\cos 2\theta})} \tag{2.68}$$

which corresponds to the polar equation of an ellipse having its center at the origin and semi-axes given by the pair $(t_+, t_-)$ shown in (2.65).

---

## 2.8 Laplace–Runge–Lenz vector

While the angular momentum and energy are natural candidates for conserved quantities in a Keplerian system, the latter also incorporates an additional hidden symmetry that goes by the name of Laplace–Runge–Lenz (LRL) vector. An interesting history on the LRL vector can be found in the papers of Goldstein and Leach and Flessas.[6] Here we simply note that the LRL vector came into notability from Pauli's work[7] on hydrogen atoms in which such a vector was exploited to derive the latter's energy levels employing certain quantum mechanical calculations. Pauli used Lenz's name in such a context although it must be said, as Goldstein[8] pointed out, that the presence of LRL-vector was implicit in the 1845 landmark paper of Hamilton entitled *Applications of Quaternions to Some Dynamical Quantum.*

Since $\vec{l}$ is a constant of motion, i.e., $\frac{d\vec{l}}{dt} = 0$, we can express the product $m\ddot{\vec{r}} \times \vec{l}$ as

$$m\ddot{\vec{r}} \times \vec{l} = \frac{d}{dt}(\vec{p} \times \vec{l}), \quad \vec{p} = m\dot{\vec{r}} \tag{2.69}$$

Thus from the radial equation of motion $m\ddot{\vec{r}} = F(r)\frac{\vec{r}}{r}$ and using $\vec{l} = \vec{r} \times \vec{p}$ we find

$$\frac{d}{dt}(\vec{p} \times \vec{l}) = F(r)\frac{\vec{r}}{r} \times (\vec{r} \times \vec{p}) \tag{2.70}$$

Now using the standard formula for a vector triple product the right-hand side turns out to be

$$= m\frac{F(r)}{r}[\vec{r}(\vec{r} \cdot \dot{\vec{r}}) - r^2\dot{\vec{r}}] \tag{2.71}$$

---

[6] P.G.L. Leach and G.P. Flessas, Generalizations of Laplace–Runge–Lenz vector, *J. Non. Math. Phys.* 10(2003) 340.

[7] W. Pauli, Uber das Wasserstoffspektrum vom Standpunkt der neuen Quantenmechanik, *Z. Phys.* 36 (1926) 336.

[8] H. Goldstein, More on the prehistory of the Laplace or RungeLenz Vector, *Am. J. Phys.* *44* (1976) 1123.

Now the product $\vec{r} \cdot \dot{\vec{r}}$ being simply $r\dot{r}$, we obtain

$$\frac{d}{dt}(\vec{p} \times \vec{l}) = mF(r)[\dot{r}\vec{r} - r\dot{\vec{r}}] \tag{2.72}$$

where unfortunately the right-hand side cannot be written as a total derivative for a general form of $F(r)$. The best we can achieve is to express (2.72) as

$$\frac{d}{dt}(\vec{p} \times \vec{l}) = -mF(r)r^2 \frac{d}{dt}(\frac{\vec{r}}{r}) \tag{2.73}$$

The non-integrability is evidently due to the presence of an arbitrary force term $F(r)$ in the right-hand side of (2.73). However, for the specific case of a repulsive force which varies inversely as the square of the distance as given by $F(r) = -\frac{\gamma}{r^2}$, the system is rendered solvable. In fact, the above equation reduces to a total derivative form, namely,

$$\frac{d}{dt}(\vec{p} \times \vec{l} - m\gamma\frac{\vec{r}}{r}) = 0 \tag{2.74}$$

implying conservation of a vector $\vec{A}$ defined by

$$\vec{A} = \vec{p} \times \vec{l} - m\gamma\hat{\vec{r}} \tag{2.75}$$

$\vec{A}$ is called the *LRL* vector. In particular if $\vec{l}$ points to the $z$-direction designated as $l_z$, then the components of $\vec{A}$ in the $xy$-plane are

$$A_x = p_y l_z - m\gamma\frac{x}{r}, \quad A_y = p_x l_z - m\gamma\frac{y}{r} \tag{2.76}$$

where $p_x$ and $p_y$ are the components of the momentum in the $xy$-plane.

A couple of observations are in order:

(i) From the form of $\vec{A}$ it follows that $\vec{A} \cdot \vec{l} = 0$. Hence $\vec{A}$ is orthogonal to $\vec{l}$ pointing that $\vec{l}$ lies in the orbital plane.

(ii) We can express the energy E explicitly in terms of the magnitudes of the $\vec{A}$ and $\vec{l}$ vectors. To this end we square the expression of $\vec{A}$ to write

$$A^2 = |\vec{p} \times \vec{l} - m\gamma\hat{\vec{r}}|^2 = p^2 l^2 + m^2\gamma^2 - 2m\frac{\gamma}{r}l^2 \tag{2.77}$$

where we have used the vector identities $(\vec{a}\times\vec{b})\cdot(\vec{c}\times\vec{d}) = (\vec{b}\cdot\vec{d})(\vec{a}\cdot\vec{c}) - (\vec{b}\cdot\vec{c})(\vec{a}\cdot\vec{d})$ and $\vec{a} \cdot (\vec{b} \times \vec{c}) = (\vec{a} \times \vec{b}) \cdot \vec{c}$. Noticing that for our inverse square law for the force implies a potential $V(r) = -\frac{\gamma}{r}$, we have the total energy as $E = \frac{p^2}{2m} - \frac{\gamma}{r}$ which can be reexpressed in terms of $A^2$ as

$$E = \frac{A^2 - m^2\gamma^2}{2ml^2} \tag{2.78}$$

This gives the relation between the energy and the square of the magnitude of the *LRL* vector.

## 2.9 Summary

In this chapter we looked at the general class of central force problems noting the distinction between inertial and gravitational mass and the related issue of the principle of equivalence. We briefly provided a derivation of Kepler's three laws and then gave a discussion of the properties and equations of orbits. A study of the general class of power law potentials was our next point of inquiry in which we examined, in particular, the stability condition of the entire class of power law potentials and mapping of the general class of potentials. The specific cases of the Coulomb and isotropic potentials were illustrated and finally a treatment of Laplace–Runge–Lenz vector was given.

# Exercises

1. Demonstrate the equivalence of the integral (2.46) with (2.44).

2. Show from (2.43) that the case $\nu = -2$ leads to orbits for a Coulomb potential. Also address the dual case of the oscillator potential.

3. Show for the ellipse equation as written down in (2.57) one of the foci is at its origin.

4. Determine the transverse velocity of the circular orbit which results due to a balance between the centrifugal and the gravitational forces.

5. Using the vector identity $\vec{r} \cdot (\vec{p} \times \vec{l}) = (\vec{r} \times \vec{p}) \cdot \vec{l} = l^2$, show that the expression (2.75) represents a conic by taking a dot product with $\vec{r}$.

# Chapter 3

## Lagrangian formulation in mechanics



In this chapter we take up the mathematical formulation of the Lagrangian approach to classical mechanics that provides a basis for its analytical foundation. The main aim is to do away with the forces of constraint which could be present in the Newtonian equations of motion. These are unknown quantities there and require to be solved for. The presence of constraints limits the freedom of the system and reduces its degrees of freedom to describe the motion. The independent coordinates (not necessarily Cartesian) accounting for the number of the degrees of freedom are termed generalized coordinates. By invoking the principle of virtual work and making a judicious use of D'Alembert's principle, a completely new form of the equations of motion emerges which replaces the Newtonian ones by a set of equations called Lagrange's equations. In place of force and acceleration, these equations are dependent on the kinetic and potential energies of the system. In the Lagrangian scheme of things we do not have to worry about the forces of constraints. Of course, no new physics emerges but the Lagrangian approach gives a fundamental starting point toward building up of a solid analytical structure of classical mechanics which proves very powerful in applications to a wide variety of problems. In the following we begin with the constraints and the idea of generalized coordinates.

## 3.1 Constraints and generalized coordinates

For a physical system consisting of a finite number of particles moving under the action of a prescribed force, we usually find that the particles have their coordinates restricted in one way or another. We can think of numerous examples in this regard. The simplest one that comes to mind is the case of a simple plane pendulum. As already discussed in Chapter 1, such a system is composed of a bob of mass $m$ attached to a massless, inextensible string of length $l$, the other end of the string being suspended from a pivot, without friction. Its position can be described by the angle $\theta$ that the string makes with the vertical line through the support or the horizontal component of the displacement of the bob. We say that the degree of freedom for the plane pendulum is one.

There are systems that require two or more independent coordinates for the description of their motion. These are called muliti-degree-of-freedom systems. For instance, if we replace the plane pendulum by taking some flexible string or if we attach a string to a rigid body at a point different from its center of mass, then we encounter a system with two degrees of freedom. This means that we need two independent coordinates to fully understand the motion of the system.

Consider the case of a rigid body which is a material body composed of an aggregate of particles, the relative distance between every pair of points in it is invariable. A rigid body does not yield to any deformation of shape arising out of expansion or contraction although it can move from one position to another as a whole as observed by an external observer. It may be interpreted as a continuous distribution and that we can arbitrarily subdivide it into small parts thus facilitating the use of laws of motion and integral calculus to understand their changes of position. The number of degrees of freedom of a rigid body in space is estimated easily. First of all, its orientation is fully determined when any three points of it are known. Now a point in space can be represented by three coordinates implying a total of nine coordinates for these three points of the rigid body. Next, the three lines joining the three points have to have their lengths invariable since the mutual distances of the points in a rigid body remain fixed. This puts three restrictions. Hence, a rigid body can have six degrees of freedom. Looked at in a different way, the six degrees of freedom account for the fact that a rigid body can have three translations along with three rotary motions. The degree of freedom is therefore defined as the number of independent coordinates (excluding the time variable) required to completely specify the position and configuration of the system.

Let us now turn to a mechanical system composed of $N$ particles. Newton's equation of motion for the $i^{th}$ particle is given by

$$m_i \ddot{\vec{r}}_i = \vec{F}_i, \quad i = 1, 2, ..., N \tag{3.1}$$

Since each particle can be specified by three coordinates, we have at hand $3N$ coordinates which are subjected to, say, $k(\leq 3N)$ equations of constraints. The number of degrees of freedom is then given by $n = 3N - k$.

A constraint is some kind of a restriction on the motion of a particle. The forces responsible for the restriction are called the forces of constraint. The forces of constraint are initially unknown and require to be solved for. For instance, the tension of the string in the plane pendulum problem is the force of constraint which is to be determined by solving the equations of motion. If the forces other than the forces of constraint are designated as *applied forces* then $\vec{F}$ appearing in (3.1) can be split up in the manner

$$m_i\ddot{\vec{r_i}} = \vec{F_i}^{(a)} + \vec{F_i}^{(c)}, \quad i = 1, 2, ..., N \tag{3.2}$$

where $\vec{F_i}^{(a)}$ are the applied forces and $\vec{F_i}^{(c)}$ are the constraint forces.

Generally, a constraint is of the form

$$\vec{\chi}(\vec{r_j}, \vec{v_j}, t) = 0, \quad j = 1, 2, .., N \tag{3.3}$$

where $\vec{v_j} \equiv \dot{\vec{r_j}}$ are the velocities. In the absence of velocities, (3.3) is called a finite or geometric constraint:

$$\vec{\chi}(\vec{r_j}, t) = 0, \quad j = 1, 2, .., N \tag{3.4}$$

Otherwise, (3.3) is a *differential* or kinematical constraint.

A particular class of differential constraints is of the linear form

$$\sum_{j=1}^{N} \vec{a_j}.\vec{v_j} + F = 0 \tag{3.5}$$

where the coefficients $\vec{a_j}$ are not all vanishing and $F$ is a scalar function of $\vec{r}$ and $t$. In the stationary case, $\frac{\partial \vec{\chi}}{\partial t} = 0$ in (3.4) while in (3.5) $F = 0$ and $\vec{a_j}$ are functions of position only. A system is called scleronomic if it is subjected to only stationary constraints; otherwise, it is called rheonomic.

Apart from the type (3.3) which are called bilateral constraints, there can be unilateral constraints which appear as inequalities.

$$\vec{\chi}(\vec{r_j}, \vec{v_j}, t) \geq 0 \tag{3.6}$$

As an example of (3.6), one can think of a volume of gas confined in a box of lengths $a, b$ and $c$. Then the motion of the gas particles is restricted to the regions $0 \leq x \leq a, 0 \leq y \leq b$ and $0 \leq z \leq c$.

Constraints may be further distinguished into two classes. We can have the ones that are expressible as an algebraic equation involving the coordinates and time but independent of velocities. Such constraints are holonomic constraints. The corresponding system is a holonomic system. On the other hand, there can exist a larger class of constraints which are nonintegrable and certainly not reducible to the holonomic types. Such constraints are therefore nonholonomic and the system subjected to them is a nonholonomic system.

Some examples given below will help us to clarify the above issues.

**Example 3.1**

For the simple plane pendulum problem the equations of constraints are $x^2 + z^2 = l^2$ and $y = 0$ where $l$ is the length of the string and the pendulum bob is restricted to swing in the $xz-$plane.

Here the constraint is holonomic and scleronomic. If, however, the length of the pendulum changes with time due to seasonal effects, then we have a time-dependent rheonomic constraint.

**Example 3.2**

Suppose a particle moving in a three-dimensional space has its motion restricted to a circle whose radius changes with time. The constraint has the form:

$$xdx + ydy + zdz = r(t)dt$$

In general the right-hand side is not integrable in a closed form and hence it is a case of a nonholonomic constraint.

**Example 3.3**

A velocity-dependent constraint has a typical form $A\dot{x} + B\dot{y} + C\dot{z} = 0$. This of course can be reexpressed as

$$Adx + Bdy + Cdz = 0$$

If the left-hand side is an exact differential, i.e., $df$, where $f \equiv f(x, y, z)$, a simple integration converts it into a holonomic constraint. A sufficient condition for the integrability is

$$A = \frac{\partial f}{\partial x}, \quad B = \frac{\partial f}{\partial y}, \quad C = \frac{\partial f}{\partial z}$$

For the above criterion to hold the necessary and sufficient conditions are given by the set of equalities

$$\frac{\partial A}{\partial y} = \frac{\partial B}{\partial x}, \quad \frac{\partial B}{\partial z} = \frac{\partial C}{\partial y}, \quad \frac{\partial C}{\partial x} = \frac{\partial A}{\partial z}$$

Otherwise, the system under consideration is a nonholonomic one.

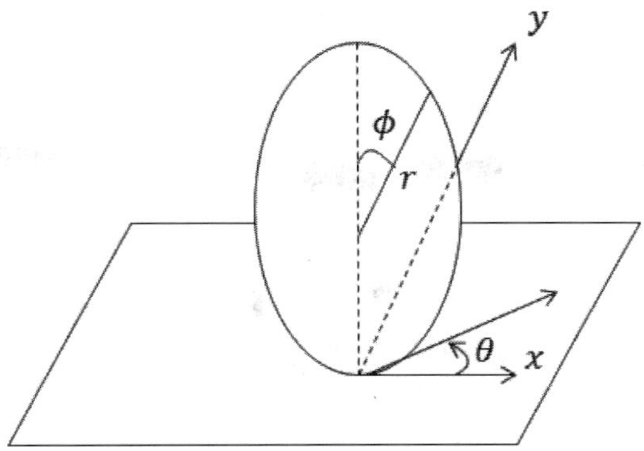


**FIGURE 3.1**: The rolling disc.

**Example 3.4**

We remarked earlier that the volume of a gas confined in a box of lengths $a, b, c$ is subjected to the unilateral constraints $0 \leq x \leq a, \quad 0 \leq y \leq b, \quad 0 \leq z \leq c$. Here the constraints are obviously nonholonomic.

**Example 3.5**

Let us consider a rolling disc problem. With the angles defined as shown in Figure 3.1, we can write down the constraint equations

$$\dot{x}\sin\theta - \dot{y}\cos\theta = 0$$
$$\dot{x}\cos\theta - \dot{y}\sin\theta = r\dot{\phi}$$

The first one speaks of no lateral motion while the second one points to pure rolling. In differential terms these are $dx = r\cos\theta d\phi, dy = r\sin\theta d\phi$ which are nonintegrable. We conclude that the problem is nonholonomic.

Once the constraint equations have been correctly identified it proves useful to set up a set of $n$ independent coordinates, $q_i (i = 1, 2, .., n)$, called *generalized coordinates*, to describe the configuration of a physical system. These generalized coordinates, whose total number equals the number of degrees of freedom available for the system, are quite general in character and need not always conform to any special type like the Cartesian or polar or say, parabolic coordinates. For the plane pendulum problem, where the constraints are $x^2 + z^2 = l^2$ and $y = 0$, any one of $x$ or $z$ or $\theta$ (the angle which the string makes with the vertical) may serve as the generalized coordinate.

There is no general rule for adopting of a particular set of generalized coordinates. As to which one needs to be employed depends a great deal upon an educated guess and also upon the conditions of a problem.

In the following section let us derive an expression for the kinetic energy in terms of the generalized coordinates $q_1, q_2, ..., q_n$.

---

## 3.2 Formulation of D'Alembert's principle

Before embarking upon the concept of virtual work, let us consider some typical situations when the forces of constraint do no work.

First consider the case of a simple plane pendulum whose length $l$ is constant. It is obvious that as it swings, the bob traces out an arc of a circle in the $xz-$plane due to the constraint $x^2 + z^2 = l^2$. The displacement of the bob is normal to the direction of the force of constraint, namely, the tension $T$, which acts along the length of the string. Hence, work done by the force of constraint is zero.

Next consider the sliding of an object down a frictionless inclined plane. Apart from the force of gravity, which is the only applied force acting upon the object, there is the normal reaction of the plane which is the force of constraint. The latter acts perpendicular to the direction toward which the object slides. It is therefore clear that such a force of constraint does no work.

Now take the case of a rigid body. Work done to move the $i^{th}$ particle of the rigid body is

$$W_i = \sum_j \vec{F_{ij}}.d\vec{r_i} \quad (i \neq j) \tag{3.7}$$

where $d\vec{r_i}$ stands for the displacement and $\vec{F_{ij}}$ represents the constraint force on the $i^{th}$ particle due to the $j^{th}$ particle and we ignore the self-forces.

To consider all the particles in the rigid body, we need to sum over $i$ in (3.7) and get for the total work done

$$W = \sum_i W_i = \sum_i \sum_j \vec{F_{ij}}.d\vec{r_i} \tag{3.8}$$

Interchanging $i$ and $j$ in (3.8) and utilizing Newton's third law,

$$\vec{F_{ij}} = -\vec{F_{ji}}$$

we arrive at the following form for $W$

$$W = \frac{1}{2} \sum_i \sum_j \vec{F_{ij}}.(d\vec{r_i} - d\vec{r_j}) \tag{3.9}$$

For a rigid body since the inter-particle distances are fixed we have $(\vec{r_i} - \vec{r_j})^2 =$ constant. Taking differential it yields $(\vec{r_i} - \vec{r_j}).(d\vec{r_i} - d\vec{r_j}) = 0$. From Newton's third law, $\vec{F_{ij}}$ acts along the relative vector $\vec{r_i} - \vec{r_j}$ which stands for the direction along the line joining the particles $i$ and $j$. It is thus implied from (3.9) that $W = 0$. Hence, the total work done by the forces of constraint in a rigid body is zero.

In all the above examples we found that the total work done by the force(s) of constraint is zero. However, the forces of constraint can do work in certain situations like the following one. Consider the case of the simple pendulum whose length $l$ is changing with time: $l = l(t)$. Here it is obvious that the bob traces out a different route than the usual circular arc. The displacement of the bob is therefore not normal to the direction of $T$, the force of constraint. So the work done by $T$ is non-zero.

How to treat to time-independent as well as time-varying constraints in a consistent framework ? Fortunately we have a way out. We invoke the concept of an infinitesimal virtual displacement which is consistent with the forces and constraints operating on the system at a given instant of time. This leads to the corresponding definition of virtual work. Here the word "virtual" has the underlying meaning that no passage of "real" time is involved during the displacements taken. In other words, we "freeze" the system at a certain point of time and think of virtual displacements $\delta\vec{r_i}(i = 1, 2, ...,)$ that are consistent with the conditions of the constraints. Some arbitrariness is, of course, involved in the choice of the direction of $\delta\vec{r_i}$. This is exploited by the principle of virtual work which we state below:

*The total virtual work done by forces of constraint $\vec{F_i}^c$ is zero for any virtual displacement.*

Mathematically it means

$$\delta W^c = \sum_{j=1}^{N} \vec{F_j^c}.\delta\vec{r_j} = 0 \tag{3.10}$$

Let us take a dot product of both sides of the equation of motion (3.2) by the virtual quantity $\delta\vec{r_i}$. It gives

$$\sum_{i=1}^{N} m_i\ddot{\vec{r_i}}.\delta\vec{r_i} = \sum_{i=1}^{N} \vec{F_i^a}.\delta\vec{r_i} + \sum_{i=1}^{N} \vec{F_i^c}.\delta\vec{r_i} \tag{3.11}$$

By virtue of the postulate (3.10) the second term in the right-hand side drops out and we are left with the sum

$$\sum_{i=1}^{N}(\vec{F_i^a} - m_i\ddot{\vec{r_i}}).\delta\vec{r_i} = 0 \tag{3.12}$$

The expression (3.12) summarizes D'Alembert's principle:[1] in it the constraint forces do not appear. Perhaps the biggest success of D'Alembert's

[1]Jean D'Alembert published in 1743 the principle that states that a combination of reversed effective forces that act on each particle of the body and external forces keep the system in equilibrium. The form as given by (3.12) was given by Joseph Lagrange later.

principle is its ability to get rid of the constraint forces. In the absence of any constraint, one can look upon $\delta\vec{r_i}$ as independent which means that their coefficients have to vanish in (3.12). The resultant equations are nothing but Newton's equations of motion for unconstrained motion. However, if the motion is a constrained one, as is indeed so in the presence of the forces of constraint, then $\delta\vec{r_i}$ are not independent and we cannot have the coefficients to be zero. We then have to deal with the single equation (3.12) whose nature can be very complicated.

As an application of D'Alembert's principle let us again focus attention on the plane pendulum problem. The generalized coordinate is chosen as $\theta$. So the work done by the applied force $mg$ is $(-mg\sin\theta)(l\delta\theta)$ where $\delta\theta$ is the virtual displacement that $\theta$ undergoes. The acceleration of the bob is $l\ddot{\theta}$. We thus have from (3.12)

$$(-mg)(l\delta\theta\sin\theta) - ml\ddot{\theta}(l\delta\theta) = 0 \tag{3.13}$$

$$\text{or} \quad \ddot{\theta} = \left(-\frac{g}{l}\right)\sin\theta = -\frac{g}{l}\theta \tag{3.14}$$

for small $\theta$. This is the usual equation of the simple harmonic motion.

Next suppose that the length of the string is changing with time. Here the only change from (3.12) is that the pendulum bob has a component of acceleration $(l\ddot{\theta} + 2\dot{l}\dot{\theta})$ in the $\theta$−direction. Hence the work done by the inertial force is $m(l\ddot{\theta} + 2\dot{l}\dot{\theta})l\delta\theta$. So (3.12) is modified to the form

$$(-mg)(l\delta\theta\sin\theta) - m(l\ddot{\theta} + 2\dot{l}\dot{\theta})(l\delta\theta) = 0$$

yielding

$$\frac{d}{dt}(ml^2\dot{\theta}) = -mgl\sin\theta \tag{3.15}$$

The left-hand side of (3.15) is the rate of change of the angular momentum of the bob about the point of support which, in the absence of gravity, remains constant even when the length of the pendulum is changing with time.

---

## 3.3 Kinetic energy of a holonomic system

A constraint-free system of $N$-particles has $3N$ coordinates and hence $3N$ degrees of freedom. Let us impose $k$ holonomic constraints upon such a system. The degrees of freedom get reduced to $3N-k$ which now stand for the number of independent coordinates. We thus represent the $3N$ coordinates as functions of $n = 3N - k$ independent generalized coordinates as follows:

$$x_i = x_i(q_1, q_2, ..., q_n; t), \quad y_i = y_i(q_1, q_2, ..., q_n; t), \quad z_i = z_i(q_1, q_2, ..., q_n; t) \tag{3.16}$$

These equations are equivalent to the vector form

$$\vec{r_j} = \vec{r_j}(q_1, q_2, ..., q_n; t), \quad j = 1, 2, ..., n \tag{3.17}$$

By definition, the kinetic energy of a system of $n$ particles is

$$T = \frac{1}{2}\sum_{j=1}^{n} m_j |\dot{\vec{r_j}}|^2 \tag{3.18}$$

where $\dot{\vec{r_j}}$ is given by

$$\dot{\vec{r_j}} = \sum \frac{\partial \vec{r_j}}{\partial q_i}\dot{q_i} + \frac{\partial \vec{r_j}}{\partial t} \tag{3.19}$$

Substituting the expression (3.19) into (3.18) we are led to the following representation for $T$

$$T = \frac{1}{2}\sum_{i,k=1}^{n} a_{ik}\dot{q_i}\dot{q_k} + \sum_{i=1}^{n} a_i\dot{q_i} + a_0 \tag{3.20}$$

where the coefficients $a_{ik}, a_i, a_0$ are

$$\begin{aligned} a_{ik} &= \sum_{j=1}^{n} m_j \frac{\partial \vec{r_j}}{\partial q_i}.\frac{\partial \vec{r_j}}{\partial q_k} \\ a_i &= \sum_{j=1}^{n} m_j \frac{\partial \vec{r_j}}{\partial q_i}.\frac{\partial \vec{r_j}}{\partial t} \\ a_0 &= \frac{1}{2}\sum_{j=1}^{n} m_j \left(\frac{\partial \vec{r_j}}{\partial t}\right)^2 \end{aligned} \tag{3.21}$$

and $i, k = 1, 2, ..., n$.

An interesting off-shot of (3.21) is that, in the scleronomic case, both the coefficients $a_i$ and $a_0$ drop out and we are left with a homogeneous function of the second degree of the generalized velocities for the kinetic energy function:

$$T = \frac{1}{2}\sum_{i,k=1}^{n} a_{ik}\dot{q_i}\dot{q_k} \equiv T_2 \quad \text{(say)} \tag{3.22}$$

Actually $T_2$ can be shown to be always degenerate:

$$det(a_{ik})_{i,k=1}^{n} \neq 0 \tag{3.23}$$

For if the above determinant were to vanish, we would be faced with a set of homogeneous linear equations of the type

$$\sum_{k=1}^{n} a_{ik}x_k = 0, \quad i = 1, 2, ..., n \tag{3.24}$$

which has a non-zero real solution. Now if we multiply the left-hand side of (3.24) by $x_i$ and sum over $i$ we would obtain

$$0 = \sum_{i,k=1}^{N} a_{ik} x_i x_k = \sum_{j=1}^{N} m_j \left( \sum_{i=1}^{n} x_i \frac{\partial \vec{r_j}}{\partial q_i} \right)^2 \tag{3.25}$$

where (3.21) has been used. Therefore we conclude that

$$\sum_{i=1}^{n} x_i \frac{\partial \vec{r_j}}{\partial q_i} = 0, \quad j = 1, 2, ..., n \tag{3.26}$$

Since $\vec{r_j} \equiv (x_j, y_j, z_j)$, (3.26) reflects that the columns of the following Jacobian matrix $J$ are linearly dependent:

$$[J] \equiv \begin{bmatrix} \frac{\partial x_1}{\partial q_1} & \frac{\partial x_1}{\partial q_2} & \cdots & \frac{\partial x_1}{\partial q_n} \\ \frac{\partial y_1}{\partial q_1} & \frac{\partial y_1}{\partial q_2} & \cdots & \frac{\partial y_1}{\partial q_n} \\ \frac{\partial z_1}{\partial q_1} & \frac{\partial z_1}{\partial q_2} & \cdots & \frac{\partial z_1}{\partial q_n} \\ \cdots & \cdots & \cdots & \cdots \\ \cdots & \cdots & \cdots & \cdots \\ \frac{\partial x_n}{\partial q_1} & \frac{\partial x_n}{\partial q_2} & \cdots & \frac{\partial x_n}{\partial q_n} \\ \frac{\partial y_n}{\partial q_1} & \frac{\partial y_n}{\partial q_2} & \cdots & \frac{\partial y_n}{\partial q_n} \\ \frac{\partial z_n}{\partial q_1} & \frac{\partial z_n}{\partial q_2} & \cdots & \frac{\partial z_n}{\partial q_n} \end{bmatrix} \tag{3.27}$$

In other words the rank $\zeta$ of $[J]$ is less than $n$. Thus we arrive at a contradiction since according to the definition of generalized coordinates we have $n$ degrees of freedom and now we find that $\zeta < n$. Hence (3.23) holds. Further $T_2 \geq 0$ since $T_2$ stands for the kinetic energy in the stationary situation with the equality sign occurring when $\dot{q_i} = 0, i = 1, 2, ..., n$.

---

## 3.4 Lagrange's equations of motion

Joseph-Louis Lagrange's work on analytical mechanics was first published in 1788. Although it appears that Lagrange's equations, which are also sometimes referred to Euler–Lagrange's equations, were known to Leonhard Euler since the 1740s, it was Lagrange who first formulated an analytical derivation of Euler's results. In this section, we are going to set up Lagrange's equations of motion (or the generalized equations of motion) for a holonomic $N-$particle system possessing $n$ degrees of freedom. A great advantage with Lagrange's approach is that the number of unknowns exactly equals the number of degrees of freedom. This is achieved by invoking D'Alembert's principle so that constraint forces are automatically done away with. Lagrangian approach opens up a new procedure for handling particle dynamics: the main difference with its Newtonian counterpart is that the energies of the system are addressed rather than the forces themselves.

Looking at the expression (3.12) of D'Alembert's principle we see that the works done by the applied forces and inertial forces under a virtual displacement $\delta\vec{r_j}(j = 1, 2, ...N)$ are respectively given by

$$\delta W^{(a)} = \sum_{j=1}^{N} \vec{F_j}^{(a)}.\delta\vec{r_j} \tag{3.28}$$

and

$$\delta W^{(in)} = \sum_{j=1}^{N} (-m_j\ddot{\vec{r_j}}).\delta\vec{r_j} \tag{3.29}$$

In a virtual variation only displacements of the coordinates are considered and time is taken to be *frozen*, i.e., no variation of $t$ is involved. We therefore have in terms of $n$ independent generalized coordinates $q_1, q_2, ..., q_n$ the relations

$$\delta\vec{r_i} = \sum_{j=1}^{N} \frac{\partial\vec{r_i}}{\partial q_j}\delta q_j, \quad i = 1, 2, ..., n \tag{3.30}$$

Plugging (3.30) into (3.28) and (3.29), $\delta W^{(a)}$ and $\delta W^{in}$ acquire the forms

$$\delta W^{(a)} = \sum_{j=1}^{n} Q_j \delta q_j \tag{3.31}$$

where

$$Q_j = \sum_{i=1}^{N} \vec{F_i}^{(a)}.\frac{\partial\vec{r_i}}{\partial q_j} \tag{3.32}$$

and

$$\delta W^{in} = \sum_{j=1}^{n}\sum_{i=1}^{N} (-m_i\ddot{\vec{r_i}}).\frac{\partial\vec{r_i}}{\partial q_j}\delta q_j \tag{3.33}$$

Since in the right-hand side of (3.31), $Q_j$ appears attached to $\delta q_j$, which are virtual displacements for generalized coordinates, one defines $Q_j$ to be the *generalized force*. However, unlike the vectorial character of the conventional force, it is scalar.

If further the system under consideration is conservative, $\vec{F_i}^a$ can be expressed as a negative gradient of some potential function, i.e., $\vec{F_i}^a = -\vec{\nabla}_i V$ and $V = V(\vec{r_1}, \vec{r_2}, ..., \vec{r_N})$. As a result we can write

$$Q_j = -\frac{\partial V}{\partial q_j} \tag{3.34}$$

In other words $Q_j$'s are derivable also from the same potential function: $V = V(q_1, q_2, ..., q_n)$.

To interpret the right-hand side of (3.33) we need to prove two lemmas:
**Lemma 1:** The kinetic energy $T$ obeys the equation

$$\frac{\partial T}{\partial \dot{q}_j} = \sum_{i=1}^{N} m_i \dot{\vec{r}_i}.\frac{\partial \vec{r}_i}{\partial q_j}, \quad j = 1, 2, ...n \tag{3.35}$$

**Proof:** It is straightforward to deduce from (3.19) that

$$\frac{\partial \dot{\vec{r}_i}}{\partial \dot{q}_j} = \frac{\partial \vec{r}_i}{\partial q_j} \tag{3.36}$$

Moreover, from the form of $T$ given by (3.18) we have

$$\frac{\partial T}{\partial \dot{q}_j} = \sum_{i=1}^{N} m_i \dot{\vec{r}_i}.\frac{\partial \dot{\vec{r}_i}}{\partial \dot{q}_j} \tag{3.37}$$

Combining (3.36) and (3.37), (3.35) follows.

**Lemma 2:** The operators $\frac{d}{dt}$ and $\frac{\partial}{\partial q_j}$ are interchangeable in the sense

$$\frac{d}{dt}\left(\frac{\partial \vec{r}_i}{\partial q_j}\right) = \frac{\partial}{\partial q_j}\left(\frac{d\vec{r}_i}{dt}\right) \tag{3.38}$$

**Proof:** The left-hand side of (3.38) can be expanded as

$$\begin{aligned}
&= \sum_{m=1}^{n} \frac{\partial^2 \vec{r}_i}{\partial q_m \partial q_j}\dot{q}_m + \frac{\partial^2 \vec{r}_i}{\partial t \partial q_j} \\
&= \frac{\partial}{\partial q_j}\left[\sum_{m=1}^{n} \frac{\partial \vec{r}_i}{\partial q_m}\dot{q}_m + \frac{\partial \vec{r}_i}{\partial t}\right] \\
&= \frac{\partial}{\partial q_j}\left(\frac{d\vec{r}_i}{dt}\right) = \text{right-hand side of (3.38)}
\end{aligned}$$

Next operating by $\frac{d}{dt}$ upon (3.35) and using (3.36) we get

$$\begin{aligned}
\frac{d}{dt}\left(\frac{\partial T}{\partial \dot{q}_j}\right) &= \sum_{i=1}^{N}\left[m_i \ddot{\vec{r}_i}.\frac{\partial \vec{r}_i}{\partial q_j} + m_i \dot{\vec{r}_i}.\frac{d}{dt}\left(\frac{\partial \vec{r}_i}{\partial q_j}\right)\right] \\
&= \sum_{i=1}^{N}\left[m_i \ddot{\vec{r}_i}.\frac{\partial \vec{r}_i}{\partial q_j} + m_i \dot{\vec{r}_i}.\frac{\partial}{\partial q_j}(\dot{\vec{r}_i})\right] \\
&= \sum_{i=1}^{N} m_i \ddot{\vec{r}_i}.\frac{\partial \vec{r}_i}{\partial q_j} + \frac{\partial}{\partial q_j}\left[\sum_{i=1}^{N}\frac{1}{2}m_i\left|\dot{\vec{r}_i}\right|^2\right] \\
&= \sum_{i=1}^{N} m_i \ddot{\vec{r}_i}.\frac{\partial \vec{r}_i}{\partial q_j} + \frac{\partial T}{\partial q_j}
\end{aligned} \tag{3.39}$$

where we have used (3.38).

Finally we make use of (3.39) in (3.33) to arrive at

$$\delta W^{(in)} = \sum_{j=1}^{n} \left[ \frac{\partial T}{\partial q_j} - \frac{d}{dt}\left( \frac{\partial T}{\partial \dot{q}_j} \right) \right] \delta q_j \tag{3.40}$$

The representations (3.31) and (3.40) enable us to restate D'Alembert's principle (3.12) in the following form:

$$\sum_{j=1}^{n} \left[ Q_j + \frac{\partial T}{\partial q_j} - \frac{d}{dt}\left( \frac{\partial T}{\partial \dot{q}_j} \right) \right] \delta q_j = 0 \tag{3.41}$$

The quantities $\delta q_j$ being arbitrary and independent, it transpires from (3.41) that the coefficients of each $\delta q_j$ must vanish separately. In consequence it must be true that

$$\frac{d}{dt}\left( \frac{\partial T}{\partial \dot{q}_j} - \frac{\partial T}{\partial q_j} \right) = Q_j, \quad j = 1, 2, ...n \tag{3.42}$$

which actually describes a set of $n$ second-order differential equations involving $n$ generalized quantities and their velocities.

For a conservative system when (3.34) holds, (3.42) can be expressed in the form

$$\frac{d}{dt}\left( \frac{\partial L}{\partial \dot{q}_j} \right) - \frac{\partial L}{\partial q_j} = 0 \tag{3.43}$$

where $L = T - V$ and $V$ is independent of the velocities $\dot{q}_1, \dot{q}_2, ..., \dot{q}_n$. $L$ is called the Lagrangian and the Equations (3.42) or (3.43) are referred to as Euler-Lagrange's equations of motion.

It is to be stressed that the unknown forces of constraint are absent from Lagrange's equations. Further, unlike Newton's equations of motion, there is no direct reference to quantities like the force or acceleration. In contrast, only a knowledge of kinetic and potential energies is required to set up $L$. However, the appearance of the generalized forces $Q_j$ in the right-hand side of Equation (3.42), which in turn are related to the applied forces $\vec{F}_i^{(a)}$, signals that the kinetic energy $T$ needs to be evaluated in an inertial frame. The reason is that $\vec{F}_i^{(a)}$ have their origins in Newton's laws, which are valid in inertial frames only.

The form (3.43) admits of the addition of a total derivative term to $L$ without affecting the equations of motion. For if we construct a new Lagrangian $L'$ from $L$ according to

$$L'(q_j, \dot{q}_j, t) = L(q_j, \dot{q}_j, t) + \frac{d\Lambda}{dt}, \quad j = 1, 2, ...n$$

where $\Lambda$ is any differentiable function of positions and time, then it is trivial to check that $\frac{d\Lambda}{dt}$ gives a vanishing contribution to the equations of motion:

$$\frac{d}{dt}\left[\frac{\partial}{\partial \dot{q}_j}\left(\frac{d\Lambda}{dt}\right)\right] = \frac{\partial}{\partial q_j}\left(\frac{d\Lambda}{dt}\right) \tag{3.44}$$

To justify (3.44) we simply have to note that $\frac{d\Lambda}{dt} = \sum_{j=1}^{n} \frac{\partial \Lambda}{\partial q_j}\dot{q}_j + \frac{\partial \Lambda}{\partial t}$ and the result immediately follows. We therefore conclude that both $L$ and $L'$ lead to the same equations of motion.

Sometimes, depending upon the nature of the problem, a more general representation of $Q_j$ than the one given in (3.34) is called for. Suppose a velocity dependent potential exists, namely $U(q_j, \dot{q}_j, t)$, such that $Q_j$'s are derivable in the manner

$$Q_j = -\left[\frac{d}{dt}\left(\frac{\partial U}{\partial \dot{q}_j}\right) - \frac{\partial U}{\partial q_j}\right], \quad j = 1, 2, ...n \tag{3.45}$$

then a similar set of equations as (3.43) follows from (3.42) if $L$ is defined according to $L = T - U$. In Chapter 7 we shall see that a velocity dependent potential has relevance in setting up of a Lagrangian for rotating frames.

---

## 3.5 Lagrange's equations for some simple systems

### 3.5.1 Plane pendulum

The plane pendulum problem has been discussed before. Treating $\theta$ to be the generalized coordinate, the kinetic and potential energies are

$$T = \frac{1}{2}ml^2\dot{\theta}^2, \quad V = -mgl\cos\theta$$

As a result the Lagrangian for the plane pendulum problem is

$$L = T - V = \frac{1}{2}ml^2\dot{\theta}^2 + mgl\cos\theta$$

Such a Lagrangian yields the partial derivatives

$$\begin{aligned}\frac{\partial L}{\partial \theta} &= -mgl\sin\theta \\ \frac{\partial L}{\partial \dot{\theta}} &= ml^2\dot{\theta}\end{aligned}$$

resulting in the following equation of motion:

$$\frac{d^2\theta}{dt^2} + \frac{g}{l}\sin\theta = 0$$

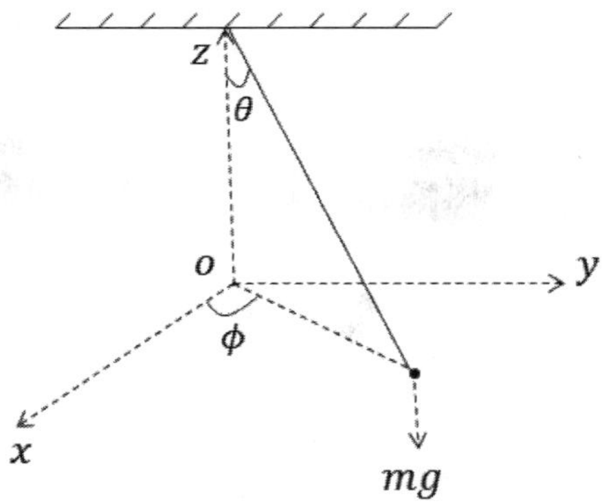


**FIGURE 3.2**: Spherical pendulum.

already referred to in Chapter 1. We therefore find that $\frac{\partial L}{\partial \dot{\theta}}$ stands for the angular momentum of the mass about the point of support and $\frac{\partial L}{\partial \theta}$ represents the torque.

We remark that $\theta$ is not the only choice of the generalized coordinate for the plane pendulum problem. We can also employ the horizontal displacement or the vertical displacement of the bob as a candidate for the generalized coordinate. However, use of the angular displacement $\theta$ appears to be the most convenient one.

### 3.5.2 Spherical pendulum

The bob of a spherical pendulum can swing in any direction in the three-dimensional space. As a result, the mass $m$ traces out a sphere of constant length $l$. See Figure 3.2. Using the polar coordinates $\theta$ and $\phi$, which work as the generalized coordinates, we find for the present problem the following forms of the kinetic and potential energies:

$$\begin{aligned} T &= \frac{1}{2}ml^2(\dot{\theta}^2 + \dot{\phi}^2 \sin^2 \theta) \\ V &= -mgl\cos\theta \end{aligned}$$

These imply for the Lagrangian

$$L = \frac{1}{2}ml^2(\dot{\theta}^2 + \dot{\phi}^2 \sin^2 \theta) + mgl\cos\theta$$

Working out the following partial derivatives corresponding to $\theta$ and $\phi$

$$\begin{aligned}
\frac{\partial L}{\partial \theta} &= ml^2 \sin\theta \cos\theta \dot{\phi}^2 - mgl \sin\theta \\
\frac{\partial L}{\partial \dot{\theta}} &= ml^2 \dot{\theta} \\
\frac{\partial L}{\partial \phi} &= 0 \\
\frac{\partial L}{\partial \dot{\phi}} &= ml^2 \sin^2\theta \dot{\phi}
\end{aligned}$$

the resulting equations of motion for $\theta$ and $\phi$ take the forms

$$\ddot{\theta} = \sin\theta \cos\theta \dot{\phi}^2 - \frac{g}{l} \sin\theta$$

$$\frac{d}{dt}(ml^2 \sin^2\theta \dot{\phi}) = 0$$

The second equation states that the component of angular momentum in the $\phi$ direction is a constant of motion. More specifically we have

$$\dot{\phi} = \frac{A}{ml^2 \sin^2\theta}$$

where $B$ is a constant.

On the other hand, in the $\theta$-equation of motion, the second term in the right-hand side represents an equivalence of the gravitational torque while the first term is a fictitious force. On integrating we get the form

$$\dot{\theta}^2 = -\frac{A^2}{m^2 l^2 \cot^2\theta + \frac{2g}{l} \cos\theta} + B$$

where $B$ is a constant.

### 3.5.3 Binary star system

Consider a binary star composed of two masses $m$ and $M$. If $O$ denotes the fixed origin and $N$ the center of mass, then according to Figure 3.3

$$\vec{r_1} = \vec{R} + \vec{r_1'}, \quad \vec{r_2} = \vec{R} + \vec{r_2'}$$

where $\vec{R}$ is the center of mass and given by

$$\vec{R} = \frac{m\vec{r_1} + M\vec{r_2}}{m + M}$$

For the present binary system, the kinetic and potential energies being

$$\begin{aligned}
T &= \frac{1}{2} m \left|\dot{\vec{r_1}}\right|^2 + \frac{1}{2} M \left|\dot{\vec{r_2}}\right|^2 \\
V &= \frac{GmM}{|\vec{r_1} - \vec{r_2}|}
\end{aligned}$$

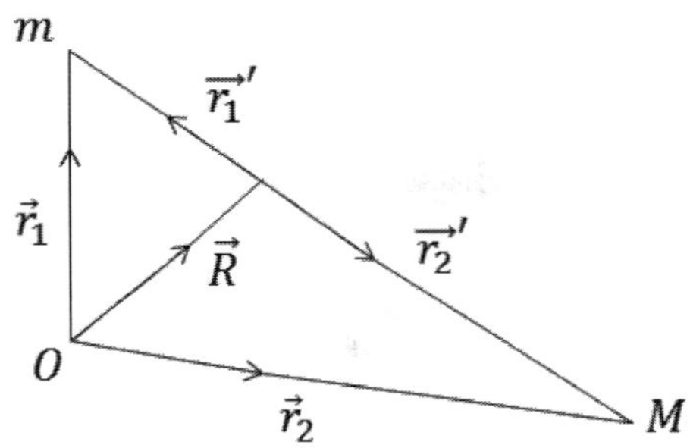


**FIGURE 3.3**: Binary star system.

the Lagrangian assumes the form

$$\begin{aligned} L &= \frac{1}{2}\left[m\left|\dot{\vec{r_1}}\right|^2 + M\left|\dot{\vec{r_2}}\right|^2\right] + \frac{GmM}{|\vec{r_1} - \vec{r_2}|} \\ &= \frac{1}{2}(m+M)\dot{\vec{R}}^2 + \frac{1}{2}\frac{mM}{m+M}\left|\dot{\vec{r}}\right|^2 + \frac{GmM}{|\vec{r}|} \end{aligned}$$

where $\vec{r} = \vec{r_1} - \vec{r_2} = \vec{r_1'} - \vec{r_2'}$.

One can see that that the coordinate $R$ does not appear explicitly implying $\frac{d}{dt}\left(\frac{\partial L}{\partial \dot{\vec{R}}}\right) = 0$. Hence the quantity $(m+M)\dot{\vec{R}}$ is a constant signifying that the kinetic energy of the system as a whole (see the first term in the Lagrangian) remains constant. Of course, such a term assumes no significance when we study the internal motion of the system.

### 3.5.4 A system with four degrees of freedom

For a system having four degrees of freedom suppose that the Lagrangian is of the type

$$L = m[\dot{q_4}^2 - \dot{q_1}^2 - \dot{q_2}^2 - \dot{q_3}^2]^{1/2} + e\sum_{k=1}^{4} A_k \dot{q_k}$$

where $A_\mu$'s are functions of coordinates alone and the parameters $m, e$ are constants.

Using the above form of $L$ we can show that the equations of motion could be given by the following sets

$$\begin{aligned} m\frac{d}{dt}(\lambda \dot{q_j}) &= e\sum_{k=1}^{4}\left(\frac{\partial A_j}{\partial q_k} - \frac{\partial A_k}{\partial q_j}\right)\dot{q_k}, \quad j = 1,2,3 \\ m\frac{d}{dt}(\lambda \dot{q_4}) &= e\sum_{k=1}^{4}\left(\frac{\partial A_k}{\partial q_4} - \frac{\partial A_4}{\partial q_k}\right)\dot{q_k} \end{aligned}$$

where $\lambda^2 = (\dot{q_4}^2 - \dot{q_1}^2 - \dot{q_2}^2 - \dot{q_3}^2)^{-1}$.

To this end, we write down for the Lagrangian $L$ the partial derivatives with respect to $\dot{q_j}$ and $q_j$

$$\begin{aligned}\frac{\partial L}{\partial \dot{q_j}} &= -m\dot{q_j}(\dot{q_4}^2 - \dot{q_1}^2 - \dot{q_2}^2 - \dot{q_3}^2)^{-1/2} + eA_j, \\ &= -m\lambda\dot{q_j} + eA_j, \\ \frac{\partial L}{\partial q_j} &= e\sum_{k=1}^{4} \frac{\partial A_k}{\partial q_j}\dot{q_k}\end{aligned}$$

where $j = 1, 2, 3$. Inserting the above expressions in Lagrange's equation of motion given by (3.43) we get

$$\frac{d}{dt}(-m\lambda\dot{q_j} + eA_j) = e\sum_{k=1}^{4} \frac{\partial A_k}{\partial q_j}\dot{q_k}, \quad j = 1, 2, 3$$

Further writing $\frac{dA_j}{dt} = \sum_{k=1}^{4} \frac{\partial A_j}{\partial q_k}\dot{q_k}$, $A_j$ being functions of position only, it follows from the above that

$$m\frac{d}{dt}(\lambda\dot{q_j}) = e\sum_{k=1}^{4}\left(\frac{\partial A_j}{\partial q_k} - \frac{\partial A_k}{\partial q_j}\right)\dot{q_k}, \quad j = 1, 2, 3$$

where $\lambda^2 = (\dot{q_4}^2 - \dot{q_1}^2 - \dot{q_2}^2 - \dot{q_3}^2)^{-1}$.

On the other hand, for $q_4$, we have

$$\begin{aligned}\frac{\partial L}{\partial \dot{q_4}} &= m\lambda\dot{q_4} + eA_4 \\ \frac{\partial L}{\partial q_4} &= \sum_{k=1}^{4} \frac{\partial A_k}{\partial q_4}\dot{q_k}\end{aligned}$$

Here writing $\frac{dA_4}{dt} = \sum_{k=1}^{4} \frac{\partial A_4}{\partial q_k}\dot{q_k}$ and using Lagrange's equation of motion for $q_4$ we find the expression

$$m\frac{d}{dt}(\lambda\dot{q_4}) = e\sum_{k=1}^{4}\left(\frac{\partial A_k}{\partial q_4} - \frac{\partial A_4}{\partial q_k}\right)\dot{q_k}$$

.

### 3.5.5 The problem of a damped oscillator

Let us consider the Lagrangian

$$L = \frac{m}{2}e^{\lambda\omega_0 t}(\dot{q}^2 - \omega_0^2 q^2), \quad \lambda < 2$$

where $m, \lambda, \omega_0$ are all positive constants. Noting that $\frac{\partial L}{\partial \dot{q}} = m\dot{q}e^{\lambda\omega_0 t}$ and $\frac{\partial L}{\partial q} = -m\omega_0^2 q e^{\lambda\omega_0 t}$, Lagrange's equation of motion is easily found to be

$$\ddot{q} + \omega_0^2 q + \lambda\omega_0\dot{q} = 0$$

As discussed in Chapter 1, the above equation represents the one for a damped oscillator. If we set $q = e^{-\frac{\lambda\omega_0}{2}t}z$, it can be transformed to the harmonic oscillator form $\ddot{z} + \bar{\omega}^2 z = 0$ where $\bar{\omega}^2 = (1 - \frac{\lambda^2}{4})\omega_0^2 > 0$ for $\lambda < 2$.

It needs to be pointed out that just by adding a damping term $m\lambda\omega_0\dot{q}$ to the Lagrangian of the harmonic oscillator, namely, $L_{HO} = \frac{m}{2}(\dot{q}^2 - \omega_0^2 q^2)$ will not produce the equation of motion for the damped oscillator. The reason is that the additional term $m\lambda\omega_0\dot{q}$ is just a total time-derivative of the quantity $\frac{m}{2}\lambda\omega_0\dot{q}^2$ and hence leads to the same equation of motion. The correct Lagrangian must contain an explicit time-dependence in an overall exponential factor as given above.

### 3.5.6 A conservative scleronomic system

For a conservative and scleronomic system we can write

$$\begin{aligned}
&\frac{d}{dt}\left[\sum_{j=1}^{n} \dot{q}_j \frac{\partial L}{\partial \dot{q}_j} - L\right] \\
&= \sum_{j=1}^{n}\left[\ddot{q}_j \frac{\partial L}{\partial \dot{q}_j} + \dot{q}_j \frac{d}{dt}\left(\frac{\partial L}{\partial \dot{q}_j}\right)\right] - \sum_{j=1}^{n}\left[\frac{\partial L}{\partial \dot{q}_j}\ddot{q}_j + \frac{\partial L}{\partial q_j}\dot{q}_j\right] \\
&= 0
\end{aligned} \tag{3.46}$$

where we have exploited Lagrange's equations $\frac{d}{dt}\left(\frac{\partial L}{\partial \dot{q}_j}\right) = \frac{\partial L}{\partial q_j}$ and noted that for a scleronomic system $\frac{\partial L}{\partial t} = 0$. Hence we conclude that

$$\sum_{j=1}^{n} \dot{q}_j \frac{\partial L}{\partial \dot{q}_j} - L = \text{constant of motion} \tag{3.47}$$

Next we may express

$$\begin{aligned}
&\sum_{j=1}^{n} \dot{q}_j \frac{\partial L}{\partial \dot{q}_j} - L \\
&= \sum_{j=1}^{n} \dot{q}_j \frac{\partial T}{\partial \dot{q}_j} - (T - V) \quad \left(\because \frac{\partial V}{\partial \dot{q}_j} = 0 \quad \text{for a conservative system}\right) \\
&= 2T - (T - V) \quad \text{(by Euler's theorem of homogeneous functions)} \\
&= T + V
\end{aligned} \tag{3.48}$$

Combining above with (3.47) we arrive at the result that for a conservative, scleronomic system the quantity $\sum_{j=1}^{n} \dot{q}_j \frac{\partial L}{\partial \dot{q}_j} - L$ is a constant of motion and represents the total energy $T + V$ is a constant.

---

## 3.6 Ignorable coordinates: Routh's procedure of solution

It often happens that for a system described by a set of $n$ generalized coordinates, a few of them say, $q_1, q_2, ..., q_k$, are not explicitly present in the Lagrangian $L$ although the corresponding velocities $\dot{q}_1, \dot{q}_2, ..., \dot{q}_k$ appear in it. Such coordinates which are absent from $L$ are called ignorable (or cyclic) coordinates. We have already encountered ignorable coordinates in some of the problems we have come across: for example, in the central force problem $\theta$ is the ignorable coordinate, in the spherical pendulum $\phi$ is absent from the Lagrangian and as such it is ignorable while for the binary star problem the coordinate $R$ does not appear explicitly in the Lagrangian. It should be clear that corresponding to any ignorable coordinate the corresponding Lagrange's equation reveals an associated constant of motion. Our task here would be to set up a modified Lagrangian addressing the remaining coordinates $q_{k+1}, q_{k+2}, ..., q_n$ from which the equations of motion can be derived. The modified Lagrangian for the explicit coordinates present in the Lagrangian is called the Routhian.

We first of all observe that for the ignorable coordinates $q_1, q_2, ..., q_k$ Lagrange's equations are

$$\frac{d}{dt}\left(\frac{\partial L}{\partial \dot{q}_r}\right) = 0, \quad r = 1, 2, ...k \tag{3.49}$$

Then the $k$ first integrals are given by

$$\frac{\partial L}{\partial \dot{q}_r} = \beta_r, \quad r = 1, 2, ...k \tag{3.50}$$

where $\beta_1, \beta_2, ..., \beta_k$ are arbitrary constants of integration.

Define now the quantity

$$R \equiv L - \sum_{r=1}^{k} \dot{q}_r \frac{\partial L}{\partial \dot{q}_r} \tag{3.51}$$

called the Routhian. Noting that the equations (3.50) are linear in $\dot{q}_1, \dot{q}_2, ...\dot{q}_k$, we can always express $\dot{q}_1, \dot{q}_2, ...\dot{q}_k$ in terms of the coordinates $q_{k+1}, q_{k+2}, ..., q_n$, the remaining velocities $\dot{q_{k+1}}, \dot{q_{k+2}}, ...\dot{q_n}$ and the quantities $\beta_1, \beta_2, ...\beta_k$. Thus, we can write $R$ as

$$R = R[q_{k+1}, q_{k+2}, ...q_n; \dot{q_{k+1}}, \dot{q_{k+2}}, ...\dot{q_n}; \beta_1, \beta_2, ...\beta_k] \tag{3.52}$$

To proceed further we take the increment of both sides of (3.51), i.e.,

$$\delta R = \delta \left[ L - \sum_{r=1}^{k} \dot{q_r} \frac{\partial L}{\partial \dot{q_r}} \right] \tag{3.53}$$

in which we notice that $L$ being ignorable in the coordinates $q_1, q_2, ..., q_k$ gives for its virtual variation

$$\delta L = \sum_{r=k+1}^{n} \frac{\partial L}{\partial q_r} \delta q_r + (\sum_{r=1}^{k} + \sum_{r=k+1}^{n}) \frac{\partial L}{\partial \dot{q_r}} \delta \dot{q_r} \tag{3.54}$$

where $t$ is treated as frozen because of the virtual variation. Therefore the variation of the second term in the right-hand side of (3.53) gives

$$\delta \left[ \sum_{r=1}^{k} \dot{q_r} \frac{\partial L}{\partial \dot{q_r}} \right] = \sum_{r=1}^{k} \delta \dot{q_r} \frac{\partial L}{\partial \dot{q_r}} + \sum_{r=1}^{k} \dot{q_r} \delta \beta_r \tag{3.55}$$

Combining (3.54) and (3.55) we have for $\delta R$

$$\delta R = \sum_{r=k+1}^{n} \frac{\partial L}{\partial q_r} \delta q_r + \sum_{r=k+1}^{n} \frac{\partial L}{\partial \dot{q_r}} \delta \dot{q_r} - \sum_{r=1}^{k} \dot{q_r} \delta \beta_r \tag{3.56}$$

where we have employed (3.50).

For the left-hand side of (3.53) considering $R$ to be the function of the arguments as specified in (3.52) gives

$$\delta R = \sum_{r=k+1}^{n} \frac{\partial R}{\partial q_r} \delta q_r + \sum_{r=k+1}^{n} \frac{\partial R}{\partial \dot{q_r}} \delta \dot{q_r} + \sum_{r=1}^{k} \frac{\partial R}{\partial \beta_r} \delta \beta_r \tag{3.57}$$

Comparing (3.56) and (3.57) and noting that the variations are arbitrary independent we get the following set of consistency conditions:

$$\begin{aligned} \frac{\partial L}{\partial q_r} &= \frac{\partial R}{\partial q_r}, \quad r = k+1, k+2, ...n \\ \frac{\partial L}{\partial \dot{q_r}} &= \frac{\partial R}{\partial \dot{q_r}}, \quad r = k+1, k+2, ...n \\ \dot{q_r} &= -\frac{\partial R}{\partial \beta_r}, \quad r = 1, 2, ...k \end{aligned} \tag{3.58}$$

The first two equations of (3.58) can be put together to read

$$\frac{d}{dt}\left( \frac{\partial R}{\partial \dot{q_r}} \right) = \frac{\partial R}{\partial q_r}, \quad r = k+1, k+2, ...n \tag{3.59}$$

while the last one implies

$$q_r = \int \frac{\partial R}{\partial \beta_r} dt, \quad r = 1, 2, ...k \tag{3.60}$$

We easily see from (3.59) that from a knowledge of the Routhian $R$, the coordinates $q_{k+1}, q_{k+2}, ...q_n$ can be determined in terms of $t$. Having got them, the remaining ones can be obtained from (3.60).

**Example 3.6**

In a dynamical system the kinetic and potential energies are

$$T = \frac{1}{2} \frac{\dot{q_1}^2}{a + bq_2^2} + \frac{1}{2}\dot{q_2}^2, \quad V = c + dq_2^2$$

Determine $q_1(t)$ and $q_2(t)$ by Routh's process of ignoration of coordinates.

From the above forms of $T$ and $V$, the Lagrangian reads

$$L = \frac{1}{2} \frac{\dot{q_1}^2}{a + bq_2^2} + \frac{1}{2}\dot{q_2}^2 - c - dq_2^2$$

in which $q_1$ can be recognized to be the ignorable coordinate. As such

$$\frac{\partial L}{\partial \dot{q_1}} = \frac{\dot{q_1}}{a + bq_2^2} = \beta$$

where $\beta$ is a constant.

The Routhian is given by

$$R = L - \dot{q_1} \frac{\partial L}{\partial \dot{q_1}}$$

Inserting the forms for $L$ and $\frac{\partial L}{\partial \dot{q_1}}$, $R$ takes the form

$$R = \frac{1}{2}\dot{q_2}^2 - \left(d + \frac{1}{2}b\beta^2\right) q_2^2 - c - \frac{1}{2}a\beta^2$$

which involves the coordinate $q_2$ and velocity $\dot{q_2}$ only.

For such an $R$, Equation (3.59) gives

$$\ddot{q_2} + (2d + b\beta^2)q_2 = 0$$

which has the solution

$$q_2 = A \sin\left[(2d + b\beta^2)^{1/2} t + \epsilon\right]$$

where $A$ and $\epsilon$ are the constants of integration.

The coordinate $q_1$ can be obtained from (3.60)

$$\begin{aligned} q_1 &= -\int \frac{\partial R}{\partial \beta} dt \\ &= \beta \int (a + bq_2^2) dt \end{aligned}$$

Substituting the above solution of $q_2$ the integration can be done in a straightforward way and we get for $q_1$

$$q_1 \equiv (\beta a + \frac{1}{2}\beta b A^2)t - \frac{\beta b A^2}{4(2d + b\beta^2)^{1/2}} \sin[(2d + b\beta^2)^{1/2} t + \epsilon] + B$$

where $B$ is a constant of integration.

**Example 3.7**

Solve the planetary problem by Routh's process of ignoration of coordinates.

We addressed the planetary problem in Chapter 2 in the context of the orbit equation for the inverse square law problem. To employ Routh's method we note that in this case the kinetic and potential energies are

$$\begin{aligned} T &= \frac{m}{2}(\dot{r}^2 + r^2\dot{\theta}^2) \\ V &= -\frac{\mu}{r} \end{aligned}$$

yielding for the Lagrangian

$$L = T - V = \frac{m}{2}(\dot{r}^2 + r^2\dot{\theta}^2) + \frac{\mu}{r}$$

Since $L$ does not contain $\theta$, it is an ignorable coordinate:

$$\begin{aligned} \frac{\partial L}{\partial \dot{\theta}} &= \text{constant} = l(\text{say}) \\ \text{or,} \quad ml^2\dot{\theta} &= l \end{aligned}$$

which implies the conservation of angular momentum.

We can now write down the Routhian

$$\begin{aligned} R &= L - \dot{\theta}\frac{\partial L}{\partial \dot{\theta}} \\ &= \frac{m}{2}\left[\dot{r}^2 - r^2\frac{l^2}{m^2 r^4}\right] + \frac{\mu}{r} \end{aligned}$$

which contains only $r$ and $\dot{r}$. We then have from (3.59)

$$m\ddot{r} = \frac{l^2}{mr^3} - \frac{\mu}{r^2}$$

whose first integral gives the conservation of total energy $E$ :

$$\frac{1}{2}m\dot{r}^2 + \frac{1}{2}\frac{l^2}{mr^2} + V = \text{constant} = E$$

as we have already know from the central force problem. The energy equation can be expressed in the integral form

$$t = \int_{r_0}^{r} \frac{dr}{\left[\frac{2}{m}\left(E - V - \frac{l^2}{2mr^2}\right)\right]^{1/2}}$$

where $r_0$ is the initial value of $r$.

On the other hand, Equation (3.60) gives

$$\begin{aligned} \theta &= -\int \frac{\partial R}{\partial l} dt \\ &= l\int_0^t \frac{dt}{mr^2(t)} + \theta_0 \end{aligned}$$

where we can determine $\theta$ explicitly if $r$ is known.

---

## 3.7 Liouville's class of Lagrangians

The Liouville's class of Lagrangians is the one for which the kinetic and potential energies appear in the special forms

$$\begin{aligned} T &= \frac{1}{2}[u_1(q_1) + u_2(q_2) + ... + u_n(q_n)][v_1(q_1)\dot{q_1}^2 + v_2(q_2)\dot{q_2}^2 + ... + v_n(q_n)\dot{q_n}^2], \\ V &= [w_1(q_1) + w_2(q_2) + ... + w_n(q_n)]/[u_1(q_1) + u_2(q_2) + ... + u_n(q_n)] \end{aligned} \quad (3.61)$$

where $u_r, v_r, w_r$ are functions of the coordinates $q_r$ only, $r = 1, 2, ...n$ for a system having $n$ degrees of freedom and the forces are conservative in nature which means that they are derivable from a potential function. A great advantage with the problems of Liouville's type is that these admit separation of variables and hence can be solved completely.

Let us effect a change of variables from the coordinates $q_1, q_2, ..., q_n$ to a new set of variables $Q_1, Q_2, ..., Q_n$ defined by

$$\dot{Q_1}^2 = v_1(q_1)\dot{q_1}^2, \dot{Q_2}^2 = v_2(q_2)\dot{q_2}^2, ..., \dot{Q_n}^2 = v_n(q_n)\dot{q_n}^2 \quad (3.62)$$

which in turn imply

$$dQ_i = \sqrt{v_i(q_i)}dq_i, \quad i = 1, 2, ..., n \tag{3.63}$$

These allow us, in principle, to integrate the above equations to obtain $Q_i$ as a function of $q_i$ alone for $i = 1, 2, ..., n$.

This enables us to transform $u_i(q_i)$ and $w_i(q_i)$ to their new forms in terms of $Q_i$, $i = 1, 2, ..., n$ which we specify by $U_r(Q_r)$ and $W_r(Q_r)$, respectively. Thus, $T$ and $V$ can be represented by

$$T = \frac{1}{2}U\sum_{j=1}^{n}\dot{Q}_j^{\,2} \tag{3.64}$$

$$V = \frac{W}{U} \tag{3.65}$$

where $U = \sum_{j=1}^{n} U_j(Q_j)$ and $W = \sum_{j=1}^{n} W_j(Q_j)$

In terms of the variables $Q_i$, which we look upon as a new generalized coordinate, Lagrange's equations read

$$\frac{d}{dt}\left(\frac{\partial T}{\partial \dot{Q}_i}\right) - \frac{\partial T}{\partial Q_i} = -\frac{\partial V}{\partial Q_i}$$

$$i.e. \quad \frac{d}{dt}(U\dot{Q}_i) - \frac{1}{2}\frac{\partial U}{\partial Q_i}(\sum_{j=1}^{n}\dot{Q}_j^{\,2}) = -\frac{\partial V}{\partial Q_i}, \quad i = 1, 2, ..., n \tag{3.66}$$

where we use the same notations for $T$ and $V$.

To tackle Equation (3.66) we need to multiply both sides by $U\dot{Q}_i$ and use (3.64) to get

$$\frac{d}{dt}(\frac{1}{2}U^2\dot{Q}_i^{\,2}) - T\dot{Q}_i\frac{\partial U}{\partial Q_i} + U\dot{Q}_i\frac{\partial V}{\partial Q_i} = 0 \tag{3.67}$$

We have already seen that for a conservative, scleronomic system, if the kinetic energy is a homogeneous quadratic function of velocities then the total energy is constant: $T + V = h$, $h$ is a constant. As such (3.67) can be put in the form

$$\frac{d}{dt}\left(\frac{1}{2}U^2\dot{Q}_i^{\,2}\right) - \dot{Q}_i\left[h\frac{\partial U}{\partial Q_i} - \frac{\partial}{\partial Q_i}(UV)\right] = 0$$

$$\text{i.e.} \quad \frac{d}{dt}\left(\frac{1}{2}U^2\dot{Q}_i^{\,2}\right) - \dot{Q}_i\left[h\frac{\partial U}{\partial Q_i} - \frac{\partial W}{\partial Q_i}\right] = 0$$

$$\text{or} \quad \frac{d}{dt}\left(\frac{1}{2}U^2\dot{Q}_i^{\,2}\right) - \dot{Q}_i\left[h\frac{\partial U_i}{\partial Q_i} - \frac{\partial W_i}{\partial Q_i}\right] = 0, \quad i = 1, 2, ..., n \tag{3.68}$$

Since, by using the chain rule of partial derivatives, we can express

$$\frac{dU_i}{dt} = \dot{Q}_i\frac{\partial U_i}{\partial Q_i}, \quad \frac{dW_i}{dt} = \dot{Q}_i\frac{\partial W_i}{\partial Q_i} \tag{3.69}$$

we arrive at the constants of motion

$$\frac{1}{2}U^2\dot{Q}_i^{\,2} - hU_i + W_i = C_i, \quad i = 1, 2, ..., n \tag{3.70}$$

where $C_i$'s are the constants of integration.

Denoting

$$d_i = C_i + hU_i - W_i, \quad i = 1, 2, ..., n \tag{3.71}$$

and restoring now to the old quantities $u, w_i, q_i$ the above constants of motion take the forms

$$\frac{1}{2}u^2\nu_i\dot{q}_i^{\,2} = d_i(q_i), \quad i = 1, 2, ..., n \tag{3.72}$$

In consequence we have

$$\sqrt{\frac{\nu_i(q_i)}{d_i(q_i)}}d(q_i) = \frac{\sqrt{2}}{u}dt, \quad i = 1, 2, ...n \tag{3.73}$$

or more explicitly

$$\int\sqrt{\frac{\nu_1(q_1)}{d_1(q_1)}}d(q_1) = \int\sqrt{\frac{\nu_2(q_2)}{d_2(q_2)}}d(q_2) + \beta_1 = ... = \int\sqrt{\frac{\nu_n(q_n)}{d_n(q_n)}}d(q_n) + \beta_{n-1} \tag{3.74}$$

In this way the variables become separated. Multiplying (3.74) by $u_1, u_2, ..., u_n$ for each respective value of $i = 1, 2, ...n$ and adding we get

$$\sum_{j=1}^{n}\int u_j\sqrt{\frac{\nu_j}{d_j}}dq_j = \int\sqrt{2}\frac{\sum u_j}{u}dt + c = \sqrt{2}t + c \tag{3.75}$$

where $c$ is the constant of integration.

Equations (3.74) and (3.75) provide the complete solution to Liouville's problem. Note that the complete solution is subject to $C_1 + C_2 + ... + C_n = 0$. This can be seen very easily from (3.72) which because of (3.71) can be written as

$$\frac{1}{2}u^2v_i(q_i)\dot{q}_i^{\,2} = C_i + hu_i(q_i) - w_i(q_i), \quad i = 1, 2, ..., n \tag{3.76}$$

Summing up over $i$ we get

$$\frac{1}{2}u^2\sum_{j=1}^{n}v_j(q_j)\dot{q}_j^{\,2} = \sum_{j=1}^{n}C_j + h\sum_{j=1}^{n}u_j - \sum_{j=1}^{n}w_j \tag{3.77}$$

In other words, $\sum C_r = u(T + V - h) = 0$ where we have used the definitions of $T$ and $V$ furnished by (3.61).

**Example 3.8**

Show that the dynamical system for which $2T = r_1 r_2(\dot{r_1}^2 + \dot{r_2}^2)$ and $V = \frac{1}{r_1} + \frac{1}{r_2}$ can be expressed as one of Liouville's types.

Let us put $r_1 = q_1 + q_2$ and $r_2 = q_1 - q_2$. $T$ and $V$ become

$$\begin{aligned} T &= (q_1^2 - q_2^2)(\dot{q_1}^2 + \dot{q_2}^2) \\ V &= \frac{2q_1}{q_1^2 - q_2^2} \end{aligned}$$

and the problem can be recognized to be of Liouville's type.

**Example 3.9**

A particle moves in a plane under the action of two Newtonian centers of attraction at the points $(c, 0)$ and $(-c, 0)$, the attractions being $\frac{\mu}{r^2}$ and $\frac{\mu'}{r'^2}$, respectively; $r, r'$ being the distances from $(c, 0)$ and $(-c, 0)$, respectively. Show that the problem is of Liouville's type.

Here

$$\begin{aligned} T &= \frac{1}{2}mv^2 = \frac{1}{2}(\dot{x}^2 + \dot{y}^2) \\ V &= -\frac{\mu}{r} - \frac{\mu'}{r'} = -\frac{\mu}{\sqrt{(x-c)^2 + y^2}} - \frac{\mu'}{\sqrt{(x+c)^2 + y^2}} \end{aligned}$$

Let us set $r' = q_1 + q_2$ and $r = q_1 - q_2$. Then $T$ and $V$ can be seen in the forms

$$\begin{aligned} T &= \frac{1}{2}(q_1^2 - q_2^2)\left[\frac{\dot{q_1}^2}{q_1^2 - c^2} + \frac{\dot{q_2}^2}{c^2 - q_2^2}\right] \\ V &= -\left[(\mu + \mu')\frac{q_1}{q_1^2 - q_2^2} + (\mu - \mu')\frac{q_2}{q_1^2 - q_2^2}\right] \end{aligned}$$

So the problem is Liouville's type.

**Example 3.10**

If $T = \frac{1}{2}(q_1^2 + q_2^2)(\dot{q_1}^2 + \dot{q_2}^2)$ and $V = \frac{1}{q_1^2 + q_2^2}$, solve the problem completely using Liouville's approach.

For this problem one can easily identify

$$u_1(q_1) = q_1^2, u_2(q_2) = q_2^2, v_1(q_1) = 1, v_2(q_2) = 1, w_1(q_1) = 1, w_2(q_2) = 0$$

Then from (3.74) we write

$$\int \sqrt{\frac{v_1(q_1)}{d_1(q_1)}}dq_1 = \int \sqrt{\frac{v_2(q_2)}{d_2(q_2)}}dq_2 + \beta$$

where $d_1(q_1)$ and $d_2(q_2)$ are given by

$$\begin{aligned} d_1(q_1) &= C_1 + hu_1(q_1) - w_1(q_1) = C_1 + hq_1^2 - 1 \\ d_2(q_2) &= C_2 + hu_2(q_2) - w_2(q_2) = -C_1 + hq_2^2 \quad (\because C_1 + C_2 = 0) \end{aligned}$$

$$\therefore \int \sqrt{\frac{1}{C_1 + hq_1^2 - 1}} dq_1 = \int \sqrt{\frac{1}{-C_1 + hq_2^2}} dq_2 + \beta$$

Integration gives

$$\cos^{-1} \frac{q_1}{\sqrt{\frac{1-C_1}{h}}} - \cos^{-1} \frac{q_2}{\sqrt{\frac{C_1}{h}}} = \text{constant} = \pi - C_0(say)$$

We thus arrive at the form

$$a^2 q_1^2 + b^2 q_2^2 + 2abq_1 q_2 \cos C_0 = \sin^2 C_0$$

where the quantities $a$ and $b$ stand for $a = \sqrt{\frac{h}{1-C_1}}$ and $b = \sqrt{\frac{h}{C_1}}$.

---

## 3.8 Small oscillations

In the theory of small oscillations one considers the perturbed behavior near the equilibrium configuration of a mechanical system and inquires if the system has a tendency to return to its original position given a slight disturbance from the position of equilibrium. The system is said to be in stable equilibrium if such a feature holds. The case of a suspended pendulum in which a point mass, tied to an inextensible string of negligible mass and hanging vertically down, furnishes one such example. On the other hand, if a slight disturbance causes a significant deviation of the system from its original position the system is said to be in a state of unstable equilibrium. A rod standing on its one end is an example of this type.

Consider a holonomic dynamical system with $n$ degrees of freedom. Let $q_1, q_2, \ldots, q_n$ be the generalized coordinates and the forces derivable from a potential function $V$. The equilibrium position of the system is obtained by solving the following set of equations:

$$\frac{\partial V}{\partial q_1} = \frac{\partial V}{\partial q_2} = \ldots = \frac{\partial V}{\partial q_n} = 0 \tag{3.78}$$

Let us subject the system to a small perturbation so that all the coordinates and velocities in the subsequent motion remain small. Taking, without loss of generality, the coordinates to be located at the zero position, we carry out a Taylor expansion of $V$ about the point $q_1 = q_2 = ... = q_n = 0$ to write

$$V = V_0 + \frac{\partial V}{\partial q_i}|_0 q_i + \frac{1}{2}\frac{\partial^2 V}{\partial q_i \partial q_j}|_0 q_i q_j + ... \quad i, j = 1, 2, ..., n \tag{3.79}$$

where the suffix (0) denotes that the corresponding quantity is to be determined at the equilibrium position and it is implied that repeated indices are summed over. The first term in the right-hand side is a mere constant and does not affect the physics of the problem while the second term vanishes at the equilibrium position. If we neglect third and higher order terms, $V$ reads simply

$$V = d_{ij} q_i q_j, \quad i, j = 1, 2, ..., n \tag{3.80}$$

where $d_{ij}$ are symmetric quantities and stand for the second-order derivatives $\frac{1}{2}\frac{\partial^2 V}{\partial q_i \partial q_j}|_0$ evaluated at the equilibrium point and so are constants.

With the kinetic energy being quadratic in velocities and given by the form $T = c_{ij}\dot{q}_i\dot{q}_j$, where $c_{ij}$'s $(i, j = 1, 2, ..., n)$ are symmetric and functions of the coordinates $q_1$, $q_2$,...,$q_n$ only, we can expand $T$ to write

$$T = c_{ij}|_0 \dot{q}_i \dot{q}_j + \frac{\partial c_{ij}}{\partial q_i}|_0 q_i \dot{q}_i \dot{q}_j + ... \quad i, j = 1, 2, ..., n \tag{3.81}$$

The second term corresponds to higher order terms and can be neglected. Henceforth omitting the suffix (0) we have for $T$ the term

$$T = c_{ij} \dot{q}_i \dot{q}_j, \quad i, j = 1, 2, ..., n \tag{3.82}$$

to lowest order.

The Lagrangian takes the form

$$L = c_{ij} \dot{q}_i \dot{q}_j - d_{ij} q_i q_j, \quad i, j = 1, 2, ..., n \tag{3.83}$$

which results in the equations of motion

$$c_{ij} \ddot{q}_j + d_{ij} q_j = 0, \quad i = 1, 2, ..., n \tag{3.84}$$

where the index $j$ has been summed over and the equations are linear and homogeneous with constant coefficients $c$'s and $d$'s.

Looking for solutions

$$q_i = A_i e^{\lambda t}, \quad i = 1, 2, ..., n \tag{3.85}$$

where $A_i$'s are constants which could be complex, yield the following full set of $n$ equations:

$$\begin{aligned}
(c_{11}\lambda^2 + d_{11})A_1 + (c_{12}\lambda^2 + d_{12})A_2 + ... + (c_{1n}\lambda^2 + d_{1n})A_n &= 0 \\
(c_{21}\lambda^2 + d_{21})A_1 + (c_{22}\lambda^2 + d_{22})A_2 + ... + (c_{2n}\lambda^2 + d_{2n})A_n &= 0 \\
&....\\
(c_{n1}\lambda^2 + d_{n1})A_1 + (c_{n2}\lambda^2 + d_{n2})A_2 + ... + (c_{2n}\lambda^2 + d_{nn})A_n &= 0
\end{aligned} \tag{3.86}$$

Note that the determinant has to vanish to ensure a nontrivial solution

$$\Delta \equiv \det \begin{pmatrix} c_{11}\lambda^2 + d_{11} & c_{12}\lambda^2 + d_{12} & ... & c_{1n}\lambda^2 + d_{1n} \\ c_{21}\lambda^2 + d_{21} & c_{22}\lambda^2 + d_{22} & ... & c_{2n}\lambda^2 + d_{2n} \\ ... & ... & ... & ... \\ c_{n1}\lambda^2 + d_{n1} & c_{n2}\lambda^2 + d_{n2} & ... & c_{2n}\lambda^2 + d_{nn} \end{pmatrix} = 0 \qquad (3.87)$$

Such a determinant is called a *secular* determinant and gives a polynomial in $\lambda^2$ of degree $n$.

While the kinetic energy is a positive definite function, so is $V$ since it is zero at the position of the equilibrium and we are interested in carrying out a perturbation around the stable equilibrium point. Thus seeking an oscillatory type of solution we can write

$$\lambda = -\omega^2 \qquad (3.88)$$

where $\omega^2$ is real giving the simple harmonic form

$$q_i(t) = A_i e^{i\omega t}, \quad i = 1, 2, ..., n \qquad (3.89)$$

with varying amplitudes but having the same period. The reality of $\omega^2$ can be checked by substituting the above form of $q_i$ in the equations of motion (3.84) directly, splitting $A_i$'s into real and imaginary parts and finding straightforwardly that the imaginary part vanishes. In fact, for a positive definite potential energy, $\omega$ is real too but it becomes purely imaginary for a negative definite potential energy.

Using (3.89) the equations of motion emerge as

$$\sum_{j=1}^{n} (d_{ij} - \omega^2 c_{ij}) A_j = 0, \quad i = 1, 2, ..., n \qquad (3.90)$$

where $A_j$ play the role of an eigenvector while $\Delta = 0$ stands for

$$\det(d_{ij} - \omega^2 c_{ij}) = 0, \quad i, j = 1, 2, ..., n \qquad (3.91)$$

which has $n$ positive semi-definite solutions for $\omega^2$. We label these as $\omega_i^2$, $i = 1, 2, ..., n$. Knowing these we can determine the different eigenvectors from the previous equation and normalize them appropriately.

Take two different eigenvalues labeled by $A$ and $B$. Then the corresponding equations of motion assume the form

$$\sum_{j=1}^{n} (d_{ij} - \omega_A^2 c_{ij}) A_j^{(A)} = 0, \quad \sum_{i=1}^{n} (d_{ij} - \omega_B^2 c_{ij}) A_i^{(B)} = 0 \qquad (3.92)$$

Noting that both the matrices $d_{ij}$ and $c_{ij}$ are symmetric, if we multiply from the left the first equation by $A_i^{(B)}$ and the second equation by $A_j^{(A)}$ and sum respectively over $i$ and $j$ we observe that the terms involving $d_{ij}$ drop out on subtracting leaving us with the relation

$$(\omega_A^2 - \omega_B^2)\sum_{i=1}^{n}\sum_{j=1}^{n}(A_i^{(A)}c_{ij}A_j^{(B)}) = 0 \tag{3.93}$$

Since the eigenvalues are different the double sum has to vanish. When the eigenvalues are equal the double sum can be normalized to unity implying the simple result

$$\sum_{i=1}^{n}\sum_{j=1}^{n}(A_i^{(A)}c_{ij}A_j^{(B)}) = \delta_{ij} \tag{3.94}$$

This is a relation for $c_{ij}$.

The one corresponding to $d_{ij}$ is given by

$$\sum_{i=1}^{n}\sum_{j=1}^{n}(A_i^{(A)}d_{ij}A_j^{(B)}) = \omega_B^2\delta_{ij} \tag{3.95}$$

where the equation of motion for $q_j^{(B)}$ is used in the above relation to eliminate $c_{ij}$.

Thus we have arrived at the conclusion that both $c_{ij}$ and $d_{ij}$ admit of simultaneous diagonalization. At this point we can revert to normal coordinates by defining a new set of variables $Q$'s

$$q_i = \sum_{i=1}^{n} A_i^{(A)}Q_{(A)} \tag{3.96}$$

This converts the Lagrangian to the normal form

$$L = \sum_{i=1}^{n}(\dot{Q}_A^2 - \omega_A^2 Q_A^2) \tag{3.97}$$

To put the above scheme in practice we discuss the following examples.

**Example 3.11: The double pendulum**

The double pendulum is a combination of two simple pendulums positioned in the same vertical plane and connected in such a way that the lower one hangs from the bob of the upper one. The bob of the upper one of mass $m_1$ is suspended from a ceiling by an inextensible string of negligible mass having a length $l_1$ while the lower bob of mass $m_2$ is tied to the bob of the upper one by an inextensible string of negligible mass having length $l_2$. The setup is illustrated in Figure 3.4.

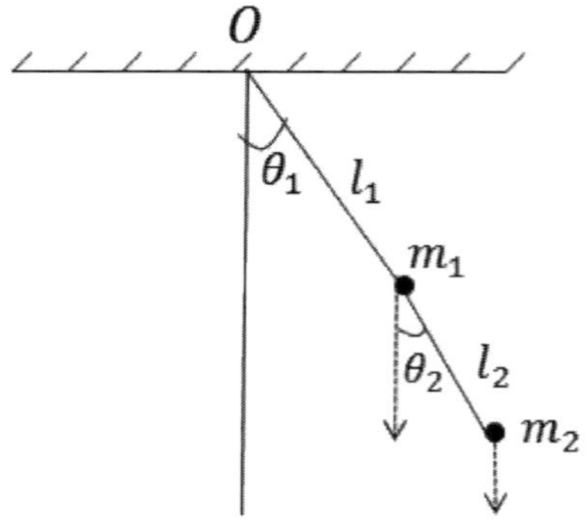


**FIGURE 3.4**: The double pendulum.

By choosing a suitable frame of reference we can introduce for the horizontal and vertical coordinates the respective relations

$$X = l_1 \sin\theta_1, \quad x = l_1 \sin\theta_1 + l_2 \sin\theta_2$$
$$Y = l_1 \cos\theta_1, \quad y = l_1 \cos\theta_1 + l_2 \cos\theta_2$$

This results in the following forms of the kinetic and potential energies

$$\begin{aligned} T &= \frac{m_1}{2}(\dot{X}^2 + \dot{Y}^2) + \frac{m_2}{2}(\dot{x}^2 + \dot{y}^2) = \frac{m_1 + m_2}{2} l_1^2 \dot{\theta_1}^2 \\ &+ \frac{m_2}{2} l_2^2 \dot{\theta_2}^2 + m_2 l_1 l_2 \cos(\theta_1 - \theta_2)\dot{\theta}_1\dot{\theta}_2 \\ V &= -m_1 g Y - m_2 g y \\ &= -(m_1 + m_2) g l_1 \cos\theta_1 - m_2 g l_2 \cos\theta_2 \end{aligned}$$

yielding for the Lagrangian

$$\begin{aligned} L = T - V &= \frac{m_1 + m_2}{2} l_1^2 \dot{\theta_1}^2 + \frac{m_2}{2} l_2^2 \dot{\theta_2}^2 + m_2 l_1 l_2 \cos(\theta_1 - \theta_2)\dot{\theta}_1\dot{\theta}_2 \\ &+ (m_1 + m_2) g l_1 \cos\theta_1 + m_2 g l_2 \cos\theta_2 \end{aligned}$$

Using Lagrange's form of the equations of motion we easily deduce the pair

$$l_1\ddot{\theta}_1 + \frac{m_2 l_2}{m_1 + m_2}\cos(\theta_1 - \theta_2)\ddot{\theta}_2 + \frac{m_2 l_2}{m_1 + m_2}\sin(\theta_1 - \theta_2)\dot{\theta_2}^2 + g\sin\theta_1 = 0$$
$$l_1 \cos(\theta_1 - \theta_2)\ddot{\theta}_1 + l_2\ddot{\theta}_2 - l_1 \sin(\theta_1 - \theta_2)\dot{\theta_1}^2 + g\sin\theta_1 = 0$$

The above equations constitute a set of nonlinear equations. For small oscillations we will be interested only in their linearized versions and as such retain terms only of small $\theta_1$ and $\theta_2$ and their time-derivatives but neglect their squares and higher powers. We are thus led to the forms

$$\ddot{\theta_1} + \frac{g}{l_1}\theta_1 = -\frac{m_2 l_2}{(m_1+m_2)l_1}\ddot{\theta_2}$$
$$\ddot{\theta_2} + \frac{g}{l_2}\theta_2 = -\frac{l_1}{l_2}\ddot{\theta_1}$$

which can be viewed as a system of two coupled differential equations.

To determine the normal modes we look into the simple case when $l_1 = l_2 = l$. The above equations then assume a greatly reduced form

$$m_l\ddot{\theta_1} + m_1 g\theta_1 = m_2 g(\theta_2 - \theta_1)$$
$$m_2 l(\ddot{\theta_1} + \ddot{\theta_2}) + m_2 g\theta_2 = 0 \tag{3.98}$$

and can be reexpressed in the manner

$$\frac{d^2}{dt^2}\begin{pmatrix}\theta_1\\ \theta_2\end{pmatrix} = M\begin{pmatrix}\theta_1\\ \theta_2\end{pmatrix}$$

where the two-dimensional matrix $M$ is given by

$$M \;=\; -\frac{g}{l}\begin{pmatrix}\frac{m_1+m_2}{m_1} & -\frac{m_2}{m_1}\\ -\frac{m_1+m_2}{m_1} & \frac{m_1+m_2}{m_1}\end{pmatrix}$$

We look for normal mode solutions for $\theta_1$ $\theta_2$

$$\begin{pmatrix}\theta_1\\ \theta_2\end{pmatrix} = e^{i\omega t}\begin{pmatrix}\bar{\theta_1}\\ \bar{\theta_2}\end{pmatrix}$$

Since the determinant of the coefficient matrix has to vanish to ensure a non-trivial solution, this means

$$\det\begin{pmatrix}-\omega^2 + \frac{g}{l}\frac{m_1+m_2}{m_1} & -\frac{g}{l}\frac{m_2}{m_1}\\ -\frac{g}{l}\frac{m_1+m_2}{m_1} & -\omega^2 + \frac{g}{l}\frac{m_1+m_2}{m_1}\end{pmatrix} = 0$$

which translates into a quadratic equation in $\omega^2$:

$$\omega^2 - \frac{g}{l}\frac{m_1+m_2}{m_1} = \pm\frac{g}{l}\sqrt{\frac{m_2(m_1+m_2)}{m_1}} = 0$$

The normal frequencies then are

$$\omega_{\pm} = \sqrt{\frac{g}{l}\frac{1}{1\mp\sqrt{\nu}}}$$

where $\nu$ stands for the ratio $\frac{m_2}{m_1+m_2}$.

Actually if we employ the normal coordinates defined by

$$\theta_\pm = \theta_1 + \nu_\pm \theta_2$$

Equation (3.98) takes the form

$$(1+\sqrt{\nu})\ddot{\theta}_+ + (1-\sqrt{\nu})\ddot{\theta}_- + \frac{g}{l}(\theta_+ + \theta_-) = 0$$
$$(1+\sqrt{\nu})\ddot{\theta}_+ + (\sqrt{\nu}-1)\ddot{\theta}_- + \frac{g}{l}(\theta_+ - \theta_-) = 0$$

and gives a pair of decoupled equations

$$\ddot{\theta_+} + \Omega_+^2 \theta_+ = 0$$
$$\ddot{\theta_-} + \Omega_-^2 \theta_- = 0$$

where the quantities $\Omega_\pm^2 = \frac{g}{l}\frac{1}{1\pm\sqrt{\nu}}$ are called the normal frequencies of oscillation.

**Example 3.12: Coupled harmonic oscillators**

We consider a system of two coupled oscillators as described by two identical harmonic oscillators that are coupled through an elastic spring. An appropriate Lagrangian is of the form

$$L = \frac{1}{2}m(\dot{x_1}^2 + \dot{x_2}^2) - \frac{1}{2}\omega_0^2(x_1^2 + x_2^2) + \gamma x_1 x_2$$

Such a system has two degrees of freedom with $x_1, x_2$ representing two generalized coordinates for its description and $\gamma$ is a coupling constant. The coordinates $x_1, x_2$ account for the displacements of the particles from their equilibrium positions.

For simplicity setting $m = 1$, the equations of motion resulting from the above Lagrangian are easily found to be

$$\ddot{x_1} - \omega_0^2 x_1 + \gamma x_2 = 0$$
$$\ddot{x_2} - \omega_0^2 x_2 + \gamma x_1 = 0$$

These equations can be put in the matrix form

$$\frac{d^2}{dt^2}\begin{pmatrix} x_1 \\ x_2 \end{pmatrix} = M \begin{pmatrix} x_1 \\ x_2 \end{pmatrix}$$

where $M$ is a $2 \times 2$ matrix given by

$$M = \begin{pmatrix} -\omega_0^2 & \gamma \\ \gamma & -\omega_0^2 \end{pmatrix}$$

As before we are concerned with the normal mode solutions like

$$\begin{pmatrix} x_1 \\ x_2 \end{pmatrix} = e^{i\omega t} \begin{pmatrix} \bar{x_1} \\ \bar{x_2} \end{pmatrix}$$

and a nontrivial solution is ensured provided

$$\det \begin{pmatrix} \omega_0^2 - \omega^2 & \gamma \\ \gamma & \omega_0^2 - \omega^2 \end{pmatrix} = 0$$

The vanishing of the determinant yields a quadratic equation $\omega$ whose eigenvalue solutions lead to the following accompanying eigenvectors:

$$\omega_1 = \sqrt{\omega_0^2 - \gamma}, \qquad \begin{pmatrix} x_1 \\ x_2 \end{pmatrix} = \frac{1}{\sqrt{2}} \begin{pmatrix} 1 \\ 1 \end{pmatrix}$$

$$\omega_2 = \sqrt{\omega_0^2 + \gamma}, \qquad \begin{pmatrix} x_1 \\ x_2 \end{pmatrix} = \frac{1}{\sqrt{2}} \begin{pmatrix} 1 \\ -1 \end{pmatrix}$$

These show a symmetric mode and an antisymmetric mode of the normal mode of oscillations. The former case reveals an oscillation in the same direction while the latter points to an out of phase oscillation. Note that if the eigenvector solutions are $\xi_1, \xi_2$ then these may be expressed in terms of $x_1, x_2$ through $\xi_1 = \frac{x_1+x_2}{\sqrt{2}}, \xi_2 = \frac{x_1-x_2}{\sqrt{2}}$.

For the weak coupling case when $\gamma << \omega_0^2$, we have the approximations

$$\omega_1 = \omega_0(1 - \frac{\gamma}{2\omega_0^2}), \quad \omega_2 = \omega_0(1 + \frac{\gamma}{2\omega_0^2})$$

Corresponding to these approximations the two normal mode solutions can be superposed to give for $x_1$ and $x_2$ the following results:

$$(x_1, x_2) = e^{\omega_0(1-\frac{\gamma}{2\omega_0^2})t} \pm e^{\omega_0(1+\frac{\gamma}{2\omega_0^2})t} \tag{3.99}$$

where we suppressed an overall constant factor. Apart from a phase term $e^{\omega_0 t}$, the two solutions consist of factors $\cos\omega' t$ and $\sin\omega' t$, where $\omega'$ stand for the ratio $\frac{\gamma}{2\omega_0}$, reflecting an out of phase behavior.

---

## 3.9 Summary

In this chapter we reviewed the principles of the Lagrangian approach to classical mechanics whose analytical structure is described by a set on $n$ ordinary second-order differential equations in terms of a Lagrangian function. After discussing the constraints, the principle of virtual work and D'Alembert's principle we derived the Lagrangian equations of motion. We then considered their applications to several simple systems. We gave a special emphasis on the role of ignorable coordinates in physical systems. Apart from offering simplification of the corresponding equation of motion ignorable coordinates furnish the important result that the generalized momenta corresponding to them are constants of motion. We described Routh's procedure of solving equations of of motion when some of the coordinates are ignorable. Subsequently we investigated Liouville's class of Lagrangians and provided a method toward their solvability. We also carried out a general treatment of the problem of small oscillations about a stable equilibrium.

# Exercises

1. The Lagrangian for a mechanical system of two degrees of freedom is given by

$$L = \frac{ma^2}{72}(16\dot{q_1}^2 + 20\dot{q_1}\dot{q_2} + 25\dot{q_2}^2) - \frac{mga}{2}(\frac{1}{3}q_1^2 + \frac{5}{6}q_2^2)$$

where $a$ is a constant. Solve the system subject to the initial conditions $\dot{q_1} = \dot{q_2} = 0$ and $q_1 = q_2 = \bar{q}$.

2. The kinetic and potential energies of a mechanical system with two degrees of freedom are $T = \frac{1}{2}(\dot{q_1}^2 + \dot{q_2}^2), V = f(q_1 - q_2)$ where $f$ is an arbitrary function of the given argument. By making a change of variables solve the problem in terms of an integral involving $f$.

3. The kinetic and potential energies of a mechanical system are in the forms $T = \sum_1^n f_i(q_i)\dot{q_i}^2, V = \sum_1^n V_i q_i$ for a system with $n$ degrees of freedom. Show that the problem can be solved by the method of quadrature.

4. A particle of mass $m$ slides down under the action of gravity on the inside of a smooth surface given by the equation $z = x^2 + y^2$. Use cylindrical coordinates and employ Routh's procedure to convert the problem to one degree of freedom.

5. The Lagrangian of a mechanical system with two degrees of freedom has the form

$$L = \frac{\dot{q_1}^2}{aq_2 + b} + \frac{1}{2}\dot{q_2}^2 + 2q_2^3 + cq_2$$

where $a, b, c$ are constants. Find an integral for $q_2$.

6. The kinetic and potential energies of a mechanical system with two degrees of freedom are $T = \frac{1}{2}\frac{\dot{q_1}^2}{a+bq_2} + \frac{1}{2}\dot{q_2}^2 q_2^2, V = c + dq_2$ where $a, b, c, d$ are constants. Show that the coordinate $q_2$ satisfies the equation $(q_2 - \mu)(q_2 + 2\mu)^2 = \nu(t - t_0)^2$, where $\mu, \nu$ are constants and $t = t_0$ is the initial value of time.

7. The kinetic and potential energies of a particle are $T = \frac{1}{2}(\dot{x}^2 + \dot{y}^2), V = -\frac{\mu}{r} - \frac{\mu'}{r'}$, where $\mu, \mu'$ are constants. Further, $r$ and $r'$ denote the respective distances of the particle from the points whose coordinates are $(a, 0)$ and $(-a, 0)$. Reduce the problem to Liouville's form.

8. A particle of mass $m$ is located in a one-dimensional potential field $V(x) = \frac{\mu}{x^2} - \frac{\nu}{x}$, where $\mu, \nu$ are positive constants. Show that the period of small oscillations that the particle performs about the equilibrium position will be $4\pi\sqrt{\frac{2m\mu^3}{\nu^4}}$.

9. Find the eigenfrequencies and eigenvectors of the normal modes for a coupled system of oscillators given by the following Lagrangian:

$$L = \frac{1}{2}(m_1\dot{x_1}^2 + m_2\dot{x_2}^2) - \frac{1}{2}(\omega_1^2 x_1^2 + \omega_2^2 x_2^2) + \gamma x_1 x_2$$

where $m_1, m_2, \omega_1, \omega_2$ are the respective masses and angular frequencies.

10. A bead of mass $m$ slides down under the action of gravity on the inside of a frictionless cyclodial wire given by the usual equation in the parametric form $x = a(\theta - \sin\theta), \quad y = a(1 + \cos\theta)$, where $0 \le \theta \le 2\pi$. Set up the Lagrangian and find the equation of motion. Writing $z = \cos(\frac{\theta}{2})$, show that the equation of motion can be reexpressed in the form $\frac{d^2 z}{dt^2} + \frac{g}{4a}z = 0$ implying that the bead oscillates with the period $2\pi\sqrt{\frac{4a}{g}}$.

# Chapter 4

## Hamiltonian and Poisson bracket



Like the Lagrangian, we can define another fundamental quantity, the Hamiltonian, associated with the evolution of a mechanical system. While in the Lagrangian formalism we are required to solve a set of $n$ second-order differential equations for a system of particles with $n$ degrees of freedom, Hamilton's equations, in essence, provide a factorized version of them in terms of $2n$ first-order partial differential equations in terms of the generalized coordinates $q_i$ and generalized or conjugate momentum $p_i$. The almost symmetrical appearance of these dynamical variables facilitates development of formal theories based on canonical transformations, action-angle variables and so on to which we shall come later. We refer to $(q_i, p_i)$ as the set of conjugate or canonical variables. The coordinates $(q_1, q_2, ..., q_n)$ and the momenta $(p_1, p_2, ..., p_n)$ constitute a $2n$-dimensional space called the phase space. Hamiltonian equations define completely the evolution of the system in phase space. The Lagrangian $L(q_1, q_2, ...q_n; \dot{q_1}, \dot{q_2}, ...\dot{q_n}; t)$, on the other hand, is interpreted in the configuration space with $q_i$ and $\dot{q_i}$ denoting, respectively, the coordinates and velocities at a specific point at time $t$.

## 4.1 The Hamiltonian

Consider the total time rate of change in the Lagrangian $L(q_i, \dot{q}_i, t), i = 1, 2, ...n$

$$\frac{dL}{dt} = \sum_{i=1}^{n} \frac{\partial L}{\partial q_i}\dot{q}_i + \sum_{i=1}^{n} \frac{\partial L}{\partial \dot{q}_i}\frac{d\dot{q}_i}{dt} + \frac{\partial L}{\partial t} \tag{4.1}$$

Since the second term in the right-hand side can be written as

$$\sum_{i=1}^{n} \frac{\partial L}{\partial \dot{q}_i}\frac{d\dot{q}_i}{dt} = \frac{d}{dt}\left(\sum_{i=1}^{n} \frac{\partial L}{\partial \dot{q}_i}\dot{q}_i\right) - \frac{d}{dt}\left(\sum_{i=1}^{n} \frac{\partial L}{\partial \dot{q}_i}\right)\dot{q}_i \tag{4.2}$$

we have

$$\frac{d}{dt}\left[\sum_{i=1}^{n} \frac{\partial L}{\partial \dot{q}_i}\dot{q}_i - L\right] = -\sum_{i=1}^{n}\left[\frac{\partial L}{\partial q_i} - \frac{d}{dt}\left(\frac{\partial L}{\partial \dot{q}_i}\right)\right]\dot{q}_i - \frac{\partial L}{\partial t} \tag{4.3}$$

If the coordinates $q_i$'s obey Lagrangian equations of motion, then the above equation reduces to

$$\frac{d}{dt}\left[\sum_{i=1}^{n} \frac{\partial L}{\partial \dot{q}_i}\dot{q}_i - L\right] = -\frac{\partial L}{\partial t} \tag{4.4}$$

We now define the generalized or conjugate momentum $p_i$ associated with the generalized coordinate $q_i$ to be

$$p_i = \frac{\partial L}{\partial \dot{q}_i}, \quad i = 1, 2, ...n \tag{4.5}$$

then (4.4) reads

$$\frac{d}{dt}\left[\sum_{i=1}^{n} p_i\dot{q}_i - L\right] = -\frac{\partial L}{\partial t} \tag{4.6}$$

Note that since the kinetic energy in terms of the Cartesian velocities is $T = \frac{1}{2}m(\dot{x}^2 + \dot{y}^2 + \dot{z}^2)$, the components of the linear momentum in terms of $T$ are simply $\frac{\partial T}{\partial \dot{x}}, \frac{\partial T}{\partial \dot{y}}$ and $\frac{\partial T}{\partial \dot{z}}$. It is therefore appropriate to view $\frac{\partial T}{\partial \dot{q}_i}$ as a kind of generalized momentum which when the potential is a function of position only represents $\frac{\partial L}{\partial \dot{q}_i}$ as in (4.5).

A word about the dimensions of the generalized momentum $p_i$. Since the generalized coordinate $q_i$ can have dimensions given in units of length or angle or whatever is found suitable, the units of the generalized or conjugate momentum are to be interpreted accordingly. Indeed we can express the units of $p_i$ and the force $F_i$ in terms of the dimensions of a generalized coordinate as in the following:

$$[p_i] = \frac{ML^2}{T}\frac{1}{[q_i]} \quad [F_i] = \frac{ML^2}{T^2}\frac{1}{[q_i]} \tag{4.7}$$

One sees that when the generalized coordinate $q_i$ has the dimension of length, $p_i$ has the dimension of momentum and $F_i$ has the dimension of force. When the generalized coordinate represents a rotation, it is dimensionless implying that $p_i$ has the dimension of the angular momentum, i.e., $(\frac{ML^2}{T})$ and $F_i$ has the dimension of torque, i.e., $(\frac{ML^2}{T^2})$.

The construction of the Hamiltonian from a given Lagrangian relies upon a Legendre transformation mapping the coordinates $\dot{q}_i$ to $p_i$. We introduce a function $H$ defined in the manner

$$H = \sum_{i=1}^{n} p_i \dot{q}_i - L \tag{4.8}$$

so that from (4.6)

$$\frac{dH}{dt} = -\frac{\partial L}{\partial t} \tag{4.9}$$

The function $H$ is known as Hamilton's function or more precisely the *Hamiltonian.* It is at once implied from (4.9) that if $t$ does not appear explicitly in $L$ then $H$ is reduced to a constant in time.

We already noted earlier that for a conservative, scleronomic system the quantity $\sum_{i=1}^{n} \dot{q}_i \frac{\partial L}{\partial \dot{q}_i} - L$ is a constant of motion and that if the kinetic energy is a homogeneous quadratic function of the velocities, the total energy is constant as well. In mathematical terms this means

$$\begin{aligned}
\sum_{i=1}^{n} p_i \dot{q}_i &= \sum_{i=1}^{n} \frac{\partial L}{\partial \dot{q}_i} \dot{q}_i \\
&= \sum_{i=1}^{n} \frac{\partial (T - V)}{\partial \dot{q}_i} \dot{q}_i \\
&= \sum_{i=1}^{n} \frac{\partial T}{\partial \dot{q}_i} \dot{q}_i \\
&= 2T
\end{aligned}$$

where $L = T - V$ and we have employed Euler's theorem in the last step. As such the Hamiltonian

$$\begin{aligned}
H &= 2T - (T - V) \\
&= T + V
\end{aligned}$$

corresponds to the total energy of the system. Sometimes $H$ is also interpreted as the *generalized energy.*

## 4.2 Hamiltonian canonical equations of motion

We can generate Hamiltonian canonical equations of motion from the definition of the Hamiltonian. First of all, taking the differential of (4.8) we get

$$\begin{aligned} dH &= \sum_{i=1}^{n}\left[\dot{q}_i dp_i + \left(p_i - \frac{\partial L}{\partial \dot{q}_i}\right) d\dot{q}_i - \frac{\partial L}{\partial q_i} dq_i\right] - \frac{\partial L}{\partial t} dt \\ &= \sum_{i=1}^{n}\left[\dot{q}_i dp_i - \frac{\partial L}{\partial q_i} dq_i\right] - \frac{\partial L}{\partial t} dt \end{aligned} \tag{4.10}$$

where we used the definition (4.5) of $p_i$.

Next, looking upon $H$ as a function of $q_i, p_i$ and $t$, i.e., $H = H(q_i, p_i, t), i = 1, 2, ..., n$, we can express $dH$ as

$$dH = \sum_{i=1}^{n}\left(\frac{\partial H}{\partial q_i} dq_i + \frac{\partial H}{\partial p_i} dp_i\right) + \frac{\partial H}{\partial t} dt \tag{4.11}$$

By comparing (4.10) and (4.11) one gets the set of consistency equations

$$\begin{aligned} \dot{q}_i &= \frac{\partial H}{\partial p_i} \\ \dot{p}_i &= \frac{\partial L}{\partial q_i} = -\frac{\partial H}{\partial q_i} \\ \frac{\partial H}{\partial t} &= -\frac{\partial L}{\partial t} \end{aligned} \tag{4.12}$$

These are Hamiltonian *canonical* equations of motion and were set up by William Rowan Hamilton in 1835. Hamilton's canonical equations govern the trajectories of particle motion in phase space. Hamilton's approach is in some sense a reformulation of Lagrange's second-order differential equations but in terms of a coupled pair of first order partial differential equations. Our reference to these equations as canonical equations arises from the fact that the investigation of motions subjected to the influence of a potential is just being reduced to the examination of simple differential equations of the form (4.12).

Interestingly the set of equations (4.12) is invariant under the replacements $q_i \to p_i$ and $p_i \to -q_i$. To make a judicious use of such equations it is necessary to know the given form of the Hamiltonian $H$. If a direct knowledge of $H$ is not known, then it may be ascertained from the Legendre transformation (4.8) by setting up the Lagrangian of the system. Note that in a scleronomic system, the equations for the $2n$ variables $(q_i, p_i)$ are subject to the pair

$$
\begin{aligned}
\dot{q}_i &\equiv \frac{dq_i}{dt} = \frac{\partial H}{\partial p_i} \\
\dot{p}_i &\equiv \frac{dp_i}{dt} = -\frac{\partial H}{\partial q_i}
\end{aligned}
\tag{4.13}
$$

Because of the above equations the time rate of the Hamiltonian simplifies to

$$
\begin{aligned}
\frac{dH}{dt} &= \sum_{i=1}^{n} \left( \frac{\partial H}{\partial q_i}\dot{q}_i + \frac{\partial H}{\partial p_i}\dot{p}_i \right) + \frac{\partial H}{\partial t} \\
&= \frac{\partial H}{\partial t}
\end{aligned}
\tag{4.14}
$$

showing that if $H$ does not depend explicitly upon $t$ then it is a constant of motion.

**Example 4.1**

If all the coordinates of a system are cyclic prove that the coordinates may be found out by integration. Prove further that if the system is scleronomous then the coordinates are linear functions of time.

Since the coordinates are cyclic, the Hamiltonian is only a function of $p_i$, $i = 1, 2, 3...$ and $t$

$$H = H(p_i, t), \quad i = 1, 2, ...n$$

and therefore

$$\dot{p}_i = -\frac{\partial H}{\partial q_i} = 0, \quad i = 1, 2, ...n$$

So we have

$$p_i = \text{constant} = \beta_i$$

where $\beta_i$ are constants. Hence in the present case we deal with a free-particle system.

Further

$$
\begin{aligned}
\dot{q}_i &= \frac{\partial H}{\partial p_i} = \phi_i(p_1, p_2, ...p_n; t) \quad \text{(say)} \\
&= \phi_i(\beta_1, \beta_2, ...\beta_n; t) \\
\therefore \quad q_i &= \int \phi_i(\beta_1, \beta_2, ...\beta_n; t)dt + \alpha_i
\end{aligned}
$$

where $\alpha_i$'s are constants of integration. Thus the coordinates can be found out by integration.

For a scleronomous system the time $t$ does not appear in $\phi_i$ yielding

$$q_i = \phi_i(\beta_1, \beta_2, ...\beta_n)t + \alpha_i, \quad i = 1, 2, ...n$$

The coordinates are therefore linear in time. This examples provides an example of an integrable system.

**Remark:**

For the kinetic energy function homogeneous in the second degree of the generalized velocities, the Lagrangian reads

$$L = \frac{1}{2}\sum_{j,k=1}^{n} a_{ik}\dot{q}_i\dot{q}_k - V(q) \tag{4.15}$$

where $q = (q_1, q_2, ...q_n)$ and the coefficient matrix $A$ is symmetric: $a_{ik} = a_{ki}$.

The canonical momenta are given by the equations $p_j = \frac{\partial L}{\partial \dot{q}_j} = \sum_{i=1}^{n} a_{ij}(q)\dot{q}_j$. In a matrix form they are $p = A\dot{q}$. Inversion leads to

$$\dot{q}_i = a_{ij}^{-1}p_j \tag{4.16}$$

where $a_{ij}^{-1}$ (in general $a_{ij}^{-1} \neq (a_{ij})^{-1}$) are the elements of the matrix $A^{-1}$ :

$$L = \frac{1}{2}\dot{q}^T Aq - V(q) \tag{4.17}$$

where $T$ is the transpose and we have suppressed the summation sign.

The Legendre transformation gives for the Hamiltonian

$$H = p^T\dot{q} - L \tag{4.18}$$

which can be expressed as

$$\begin{aligned}
H &= p^T\dot{q} - [\frac{1}{2}\dot{q}^T Aq - V] \\
&= p^T A^{-1}p - [\frac{1}{2}\dot{q}^T Aq - V] \\
&= p^T A^{-1}p - \frac{1}{2}p^T A^{-1}AA^{-1}p + V \\
&= \frac{1}{2}p^T A^{-1}p + V \\
&= T + V
\end{aligned}$$

Thus $H$ stands for the total energy.

**Example 4.2**

Solve the plane pendulum problem using the Hamiltonian approach.

As discussed in Chapter 3, the Lagrangian for such a system reads

$$L = \frac{1}{2}ml^2\dot{\theta}^2 + mgl\cos\theta$$

where $\theta$ is the only generalized coordinate. Therefore the generalized momentum corresponding to $\theta$ is

$$p_\theta = \frac{\partial L}{\partial \dot{\theta}} = ml^2\dot{\theta}$$

From the Legendre transformation we can construct the Hamiltonian as

$$H = p_\theta\dot{\theta} - L = \frac{1}{2}ml^2\dot{\theta}^2 - mgl\cos\theta$$

However it needs to be defined in terms of proper variables $\theta$ and $p_\theta$. Consequently we eliminate $\dot{\theta}$ in favor of $p_\theta$ to recast it as

$$H = \frac{p_\theta^2}{2ml^2} - mgl\cos\theta$$

where we have used $p_\theta = ml^2\dot{\theta}$. The accompanying Hamiltonian equations which follow are

$$\begin{aligned}
\frac{d\theta}{dt} &= \frac{\partial H}{\partial p_\theta} = \frac{p_\theta}{ml^2} \\
\frac{dp_\theta}{dt} &= -\frac{\partial H}{\partial \theta} = -mgl\sin\theta
\end{aligned}$$

We notice that $t$ does not appear explicitly in $H$. So $H$ is a constant of motion and leads to the energy equation

$$E = \frac{p_\theta^2}{2ml^2} - mgl\cos\theta$$

where the energy $E$ is the constant value of the Hamiltonian.

**Example 4.3**

Examine the motion of a particle sliding on a parabolic wire.

Let the parabolic wire be defined in the $xy$ plane along which the particle is sliding. We ignore any effect of friction and assume that the particle is acted upon by gravity only.

Taking $x$ to be the generalized coordinate we take the parabola to be given by an equation $y = \frac{1}{2}x^2$.

The Lagrangian is

$$\begin{aligned} L &= \frac{1}{2}m(\dot{x}^2+\dot{y}^2)-mgy \\ &= \frac{1}{2}m(1+x^2)\dot{x}^2-\frac{mg}{2}x^2 \\ \therefore\quad p &= \frac{\partial L}{\partial \dot{x}}=m(1+x^2)\dot{x} \end{aligned}$$

Hence the Hamiltonian is given by

$$H = T+V=\frac{p^2}{2m(1+x^2)}+\frac{mg}{2}x^2$$

which implies by the use of Hamiltonian equations

$$\dot{x}=\frac{p}{m(1+x^2)},\quad \dot{p}=-x\left[\frac{p^2}{m(1+x^2)}+mg\right]$$

Note that $p\neq m\dot{x}$.

**Example 4.4**

The Lagrangian for a free particle in terms of paraboloidal coordinates $(\xi,\eta,\phi)$ is

$$L=\frac{1}{2}m(\xi^2+\eta^2)(\dot{\xi}^2+\dot{\eta}^2)+\frac{1}{2}m\xi^2\eta^2\dot{\phi}^2$$

Set up the Hamiltonian.

The paraboloidal coordinates were defined in Chapter 1. The canonical momenta conjugate to the generalized coordinates $(\xi,\eta,\phi)$ are

$$\begin{aligned} p_\xi &= \frac{\partial L}{\partial \dot{\xi}}=m(\xi^2+\eta^2)\dot{\xi} \\ p_\eta &= \frac{\partial L}{\partial \dot{\eta}}=m(\xi^2+\eta^2)\dot{\eta} \\ p_\phi &= \frac{\partial L}{\partial \dot{\phi}}=m\xi^2\eta^2\dot{\phi} \end{aligned}$$

Hence the Hamiltonian has the form

$$\begin{aligned} H &= p_\xi\dot{\xi}+p_\eta\dot{\eta}+p_\phi\dot{\phi}-L \\ &= \frac{1}{2m}\left[\frac{p_\xi^2+p_\eta^2}{\xi^2+\eta^2}+\frac{p_\phi^2}{\xi^2\eta^2}\right] \end{aligned}$$

**Example 4.5**

Solve the planetary problem.

Here the kinetic and potential energies are

$$\begin{aligned} T &= \frac{1}{2}m(\dot{r}^2 + r^2\dot{\theta}^2) \\ V &= -\frac{\mu}{r} \\ \therefore \quad L &= T - V = \frac{1}{2}m(\dot{r}^2 + r^2\dot{\theta}^2) + \frac{\mu}{r} \end{aligned}$$

As such we can write for the canonical momenta

$$\begin{aligned} p_r &= \frac{\partial L}{\partial \dot{r}} = m\dot{r} \\ p_\theta &= \frac{\partial L}{\partial \dot{\theta}} = mr^2\dot{\theta} \end{aligned}$$

These give for the Hamiltonian

$$\begin{aligned} H &= T + V = \frac{1}{2}m(\dot{r}^2 + r^2\dot{\theta}^2) - \frac{\mu}{r} \\ &= \frac{1}{2m}\left(p_r^2 + \frac{p_\theta^2}{r^2}\right) - \frac{\mu}{r} \end{aligned}$$

We therefore deduce

$$\begin{aligned} \dot{r} &= \frac{\partial H}{\partial p_r} = \frac{1}{m}p_r, \quad \dot{p_r} = -\frac{\partial H}{\partial r} = \frac{p_\theta^2}{mr^3} = -\frac{\mu}{r^2} \\ \dot{\theta} &= \frac{\partial H}{\partial p_\theta} = \frac{p_\theta}{mr^2}, \quad \dot{p_\theta} = 0 \end{aligned}$$

Now $\dot{p_\theta} = 0$ implies

$$\begin{aligned} \frac{d}{dt}(mr^2\dot{\theta}) &= 0 \\ \text{or,} \quad mr^2\dot{\theta} &= \text{constant} = l(\text{say}) \end{aligned}$$

Eliminating $p_r, p_\theta$ and $\dot{\theta}$ from the above equation we arrive at the form

$$m\ddot{r} = \frac{l^2}{mr^3} - \frac{\mu}{r^2}$$

which has already been solved in Chapter 3 in connection with Routh's procedure of ignoration of coordinates.

**Example 4.6**

Write down Hamiltonian equations in spherical polar coordinates for the Lagrangian

$$L = \frac{1}{2}m\dot{\vec{r}}^2 + \frac{k}{r}$$

where $k$ is a constant.

In spherical polar coordinates the Lagrangian is of the form

$$L = \frac{1}{2}m(\dot{r}^2 + r^2\dot{\theta}^2 + r^2\sin^2\theta\dot{\phi}^2) + \frac{k}{r}$$

So the matrices $A$ and $A^{-1}$, defined earlier, are given by

$$\begin{aligned} A &= \begin{pmatrix} m & 0 & 0 \\ 0 & mr^2 & 0 \\ 0 & 0 & mr^2\sin^2\theta \end{pmatrix} \\ A^{-1} &= \begin{pmatrix} \frac{1}{m} & 0 & 0 \\ 0 & \frac{1}{mr^2} & 0 \\ 0 & 0 & \frac{1}{mr^2\sin^2\theta} \end{pmatrix} \end{aligned}$$

Consequently the Hamiltonian reads

$$H = \frac{1}{2m}\left(p_r^2 + \frac{1}{r^2}p_\theta^2 + \frac{1}{r^2\sin^2\theta}p_\phi^2\right) - \frac{k}{r}$$

These give for the Hamiltonian equations

$$\dot{p_r} = -\frac{\partial H}{\partial r}, \quad \dot{p_\theta} = -\frac{\partial H}{\partial \theta}, \quad \dot{p_\phi} = -\frac{\partial H}{\partial \phi}$$

which lead to

$$m\ddot{r} = \frac{\Lambda^2}{mr^3} - \frac{k}{r^2}, \quad \frac{d}{dt}(mr^2\dot{\theta}) = \frac{\cos\theta}{mr^2\sin^3\theta}p_\phi^2, \quad \dot{p_\phi} = 0$$

where $\Lambda^2 = p_\theta^2 + \frac{p_\phi^2}{\sin^2\theta}$ and $p_r = m\dot{r}, \quad p_\theta = mr^2\dot{\theta}, \quad p_\phi = mr^2\sin^2\theta\dot{\phi}$ from

$$\begin{pmatrix} p_r \\ p_\theta \\ p_\phi \end{pmatrix} = A \begin{pmatrix} \dot{r} \\ \dot{\theta} \\ \dot{\phi} \end{pmatrix}$$

## 4.3 Poisson bracket

Consider some function $f(q,p,t)$ of the canonical variables $q_i$ and $p_i$. Then for $i = 1, 2, ...n$

$$\begin{aligned}\frac{df}{dt} &= \sum_{i=1}^{n}\left(\frac{\partial f}{\partial q_i}\frac{dq_i}{dt} + \frac{\partial f}{\partial p_i}\frac{dp_i}{dt}\right) + \frac{\partial f}{\partial t}\\ &= \sum_{i=1}^{n}\left(\frac{\partial f}{\partial q_i}\frac{\partial H}{\partial p_i} - \frac{\partial f}{\partial p_i}\frac{\partial H}{\partial q_i}\right) + \frac{\partial f}{\partial t} \qquad (4.19)\end{aligned}$$

We introduce a quantity $\{u,v\}$ involving the functions $u = u(q_i, p_i, t)$ and $v = v(q_i, p_i, t)$ and defined by

$$\{u,v\} = \sum_{i=1}^{n}\left(\frac{\partial u}{\partial q_i}\frac{\partial v}{\partial p_i} - \frac{\partial u}{\partial p_i}\frac{\partial v}{\partial q_i}\right) \qquad (4.20)$$

A trivial consequence of (4.20) is the result

$$\{q_i, p_j\} = \delta_{ij} \qquad (4.21)$$

Further (4.19) can be expressed by the form

$$\frac{df}{dt} = \{f, H\} + \frac{\partial f}{\partial t} \qquad (4.22)$$

The above equation states that if $f$ is a constant of motion, then $\frac{\partial f}{\partial t} + \{f, H\} = 0$. Furthermore, if $f$ does not involve $t$ explicitly, then $\{f, H\} = 0$.

A first integral of a Hamiltonian system is defined to be a quantity that is a constant of the motion for the flow described by Hamilton's equations. It follows that if the quantity $X$ is a first integral of a Hamiltonian system for the Hamiltonian H, then the Poisson bracket $\{H, X\} = 0$. Evidently the Hamiltonian itself is a first integral. For a mechanical system with $n$ degrees of freedom described by the Hamiltonian $H$ we call it to be completely integrable if it has n first integrals including the Hamiltonian itself.

Along with (4.22) we have the relation

$$\begin{aligned}\{q_i, H\} &= \sum_{j=1}^{n}\left(\frac{\partial q_i}{\partial q_j}\frac{\partial H}{\partial p_j} - \frac{\partial q_i}{\partial p_j}\frac{\partial H}{\partial q_j}\right)\\ &= \sum_{j=1}^{n}\delta_{ij}\frac{\partial H}{\partial p_j} = \frac{\partial H}{\partial p_i} = \dot{q}_i \qquad (4.23)\end{aligned}$$

and

$$\{p_i, H\} = \dot{p}_i \qquad (4.24)$$

As a result Hamiltonian equations assume the forms

$$\begin{aligned}\frac{dq_i}{dt} &= \{q_i, H\} \\ \frac{dp_i}{dt} &= \{p_i, H\} \qquad (4.25)\end{aligned}$$

where $i = 1, 2, ...n$. The quantity $\{u, v\}$ defined by (4.20) is called the Poisson bracket of two dynamical variables $u(q_i, p_i, t)$ and $v(q_i, p_i, t)$ and plays an important role in Hamiltonian mechanics.

---

## 4.4 Properties of Poisson bracket

The Poisson bracket satisfies several interesting properties. Below we give some of these with proofs.

1. Linearity: $\{u_1 + u_2, v\} = \{u_1, v\} + \{u_2, v\}$
$\{cu, v\} = C\{u, v\}, \quad c$ a constant
2. Antisymmetry: $\{u, v\} = -\{v, u\}$
3. Product rule: $\{u, vw\} = \{u, v\}w + v\{u, w\}$
4. Jacobi identity: $\{u, \{v, w\}\} + \{v, \{w, u\}\} + \{w, \{u, v\}\} = 0$

**Proofs:**

(1)

$$\begin{aligned}\{u_1 + u_2, v\} &= \sum_{i=1}^{n}\left[\frac{\partial(u_1+u_2)}{\partial q_i}\frac{\partial v}{\partial p_i} - \frac{\partial(u_1+u_2)}{\partial p_i}\frac{\partial v}{\partial q_i}\right] \\ &= \sum_{i=1}^{n}\left[\left(\frac{\partial u_1}{\partial q_i} + \frac{\partial u_2}{\partial q_i}\right)\frac{\partial v}{\partial p_i} - \left(\frac{\partial u_1}{\partial p_i} + \frac{\partial u_2}{\partial p_i}\right)\frac{\partial v}{\partial q_i}\right] \\ &= \sum_{i=1}^{n}\left[\left(\frac{\partial u_1}{\partial q_i}\frac{\partial v}{\partial p_i} - \frac{\partial u_1}{\partial p_i}\frac{\partial v}{\partial q_i}\right)\right] + \sum_{i=1}^{n}[1 \to 2] \\ &= \{u_1, v\} + \{u_2, v\} \\ \{cu, v\} &= \sum_{i=1}^{n}\left[\frac{\partial(cu)}{\partial q_i}\frac{\partial v}{\partial p_i} - \frac{\partial(cu)}{\partial p_i}\frac{\partial v}{\partial q_i}\right] \\ &= \sum_{i=1}^{n}\left[c\frac{\partial(u)}{\partial q_i}\frac{\partial v}{\partial p_i} - c\frac{\partial(u)}{\partial p_i}\frac{\partial v}{\partial q_i}\right] \\ &= c\{u, v\}\end{aligned}$$

$$(2)\quad \{u,v\} = \sum_{i=-1}^{n}\left(\frac{\partial u}{\partial q_i}\frac{\partial v}{\partial p_i} - \frac{\partial u}{\partial p_i}\frac{\partial v}{\partial q_i}\right)$$
$$= -\sum_{i=1}^{n}\left(\frac{\partial v}{\partial q_i}\frac{\partial u}{\partial p_i} - \frac{\partial v}{\partial p_i}\frac{\partial u}{\partial q_i}\right)$$
$$= -\{v,u\}$$

$$(3)\quad \{u,vw\} = \sum_{i=1}^{n}\left[\frac{\partial u}{\partial q_i}\frac{\partial (vw)}{\partial p_i} - \frac{\partial u}{\partial p_i}\frac{\partial (vw)}{\partial q_i}\right]$$
$$= \sum_{i=1}^{n}\left[\frac{\partial u}{\partial q_i}\left(\frac{\partial v}{\partial p_i}w + v\frac{\partial w}{\partial p_i}\right) - \frac{\partial u}{\partial p_i}\left(\frac{\partial v}{\partial q_i}w + v\frac{\partial w}{\partial q_i}\right)\right]$$
$$= \sum_{i=1}^{n}\left(\frac{\partial u}{\partial q_i}\frac{\partial v}{\partial p_i} - \frac{\partial u}{\partial p_i}\frac{\partial v}{\partial q_i}\right)w + v\sum_{i=1}^{n}\left(\frac{\partial u}{\partial q_i}\frac{\partial w}{\partial p_i} - \frac{\partial u}{\partial p_i}\frac{\partial w}{\partial q_i}\right)$$
$$= \{u,v\}w + v\{u,w\}$$

$$(4)\quad \{u,\{v,w\}\} + \{v,\{w,u\}\}$$
$$= \{u,\{v,w\}\} - \{v,\{u,w\}\}$$
$$= \sum_{i=1}^{n}\left[\left\{u, \frac{\partial v}{\partial q_i}\frac{\partial w}{\partial p_i} - \frac{\partial v}{\partial p_i}\frac{\partial w}{\partial q_i}\right\} - \left\{v, \frac{\partial u}{\partial q_i}\frac{\partial w}{\partial p_i} - \frac{\partial u}{\partial p_i}\frac{\partial w}{\partial q_i}\right\}\right]$$
$$= \sum_{i=1}^{n}\left[\left\{u, \frac{\partial v}{\partial q_i}\frac{\partial w}{\partial p_i}\right\} - \left\{u, \frac{\partial w}{\partial q_i}\frac{\partial v}{\partial p_i}\right\} - \left\{v, \frac{\partial u}{\partial q_i}\frac{\partial w}{\partial p_i}\right\} + \left\{v, \frac{\partial w}{\partial q_i}\frac{\partial u}{\partial p_i}\right\}\right]$$
$$= \sum_{i=1}^{n}\left[\left\{u, \frac{\partial v}{\partial q_i}\right\}\frac{\partial w}{\partial p_i} + \frac{\partial v}{\partial q_i}\left\{u, \frac{\partial w}{\partial p_i}\right\} - \left\{u, \frac{\partial w}{\partial q_i}\right\}\frac{\partial v}{\partial p_i} - \frac{\partial w}{\partial q_i}\left\{u, \frac{\partial v}{\partial p_i}\right\}\right]$$
$$+\sum_{i=1}^{n}\left[-\left\{v, \frac{\partial u}{\partial q_i}\right\}\frac{\partial w}{\partial p_i} - \frac{\partial u}{\partial q_i}\left\{v, \frac{\partial w}{\partial p_i}\right\} + \left\{v, \frac{\partial w}{\partial q_i}\right\}\frac{\partial u}{\partial p_i} + \frac{\partial w}{\partial q_i}\left\{v, \frac{\partial u}{\partial p_i}\right\}\right]$$
$$= X + Y(\text{say})$$

where the product rule is employed and $X, Y$ defined by

$$X = \sum_{i=1}^{n}\left[\left\{u, \frac{\partial v}{\partial q_i}\right\}\frac{\partial w}{\partial p_i} - \frac{\partial w}{\partial q_i}\left\{u, \frac{\partial v}{\partial p_i}\right\} - \left\{v, \frac{\partial u}{\partial q_i}\right\}\frac{\partial w}{\partial p_i} + \frac{\partial w}{\partial q_i}\left\{v, \frac{\partial u}{\partial p_i}\right\}\right]$$
$$Y = \sum_{i=1}^{n}\left[\frac{\partial v}{\partial q_i}\left\{u, \frac{\partial w}{\partial p_i}\right\} - \left\{u, \frac{\partial w}{\partial q_i}\right\}\frac{\partial v}{\partial p_i} - \frac{\partial u}{\partial q_i}\left\{v, \frac{\partial w}{\partial p_i}\right\} + \left\{v, \frac{\partial w}{\partial q_i}\right\}\frac{\partial u}{\partial p_i}\right]$$

Now

$$\begin{aligned}
X &= \sum_{i=1}^{n}\left[\left(\left\{u,\frac{\partial v}{\partial q_i}\right\}-\left\{v,\frac{\partial u}{\partial q_i}\right\}\right)\frac{\partial w}{\partial p_i}-\frac{\partial w}{\partial q_i}\left(\left\{u,\frac{\partial v}{\partial p_i}\right\}-\left\{v,\frac{\partial u}{\partial p_i}\right\}\right)\right]\\
&= \sum_{i=1}^{n}\left[\left(\left\{u,\frac{\partial v}{\partial q_i}\right\}+\left\{\frac{\partial u}{\partial q_i},v\right\}\right)\frac{\partial w}{\partial p_i}-\frac{\partial w}{\partial q_i}\left(\left\{u,\frac{\partial v}{\partial p_i}\right\}+\left\{\frac{\partial u}{\partial p_i},v\right\}\right)\right]\\
&= \sum_{i=1}^{n}\left[\frac{\partial}{\partial q_i}\{u,v\}\frac{\partial w}{\partial p_i}-\frac{\partial w}{\partial q_i}.\frac{\partial w}{\partial p_i}\{u,v\}\right]\\
&= \{\{u,v\},w\}
\end{aligned}$$

where we have used the results

$$\begin{aligned}
\frac{\partial}{\partial q_i}\{u,v\} &= \left\{u,\frac{\partial v}{\partial q_i}\right\}+\left\{\frac{\partial u}{\partial q_i},v\right\}\\
\frac{\partial}{\partial p_i}\{u,v\} &= \left\{u,\frac{\partial v}{\partial p_i}\right\}+\left\{\frac{\partial u}{\partial p_i},v\right\}
\end{aligned}$$

which can be proved easily.

We next show that $Y=0$. Indeed it is straightforward to see on expanding the Poisson brackets that

$$\begin{aligned}
Y &= \sum_i\sum_j[\frac{\partial^2 w}{\partial p_i\partial p_j}\left(\frac{\partial u}{\partial q_i}\frac{\partial u}{\partial q_j}-\frac{\partial u}{\partial q_i}\frac{\partial u}{\partial q_j}\right)+\frac{\partial^2 w}{\partial q_i\partial q_j}\left(\frac{\partial u}{\partial p_j}\frac{\partial u}{\partial p_i}-\frac{\partial u}{\partial p_i}\frac{\partial v}{\partial p_j}\right)\\
&\quad+\frac{\partial^2 w}{\partial q_j\partial p_i}\left(-\frac{\partial v}{\partial q_i}\frac{\partial u}{\partial p_j}+\frac{\partial u}{\partial q_i}\frac{\partial u}{\partial p_j}\right)+\frac{\partial^2 w}{\partial q_i\partial p_j}\left(-\frac{\partial u}{\partial q_j}\frac{\partial u}{\partial p_i}+\frac{\partial v}{\partial q_j}\frac{\partial u}{\partial p_i}\right)]
\end{aligned}$$

On interchanging the indices $i$ and $j$ the first and second terms vanish due to antisymmetry while the third term becomes equal to the fourth term except for a sign. We therefore conclude that $Y=0$.

Hence

$$\{u,\{v,w\}\}+\{v,\{w,u\}\}=X=\{\{u,v\},w\}$$

which stands for the Jacobi identity.

---

## 4.5 Poisson theorem

If corresponding to a given holonomic system $u$ and $v$ are two constants of motion, then their Poisson bracket is also a constant of motion.

**Proof:**

Using (4.20) we can write

$$\frac{d}{dt}\{u,v\} = \{\{u,v\},H\} + \frac{\partial}{\partial t}\{u,v\}$$

Using the product rule and Jacobi identity we obtain on rearrangement

$$\begin{aligned}
\frac{d}{dt}\{u,v\} &= \left\{u, \frac{\partial u}{\partial t}\right\} + \left\{\frac{\partial u}{\partial t}, v\right\} - \{\{v,H\},u\} - \{\{H,u\},v\} \\
&= \left\{\frac{\partial u}{\partial t} + \{u,H\}, v\right\} + \left\{u, \frac{\partial v}{\partial t} + \{v,H\}\right\} \\
&= \left\{\frac{du}{dt}, v\right\} + \left\{v, \frac{dv}{dt}\right\} \\
&= 0
\end{aligned} \tag{4.26}$$

since $\frac{du}{dt} = 0$, $\frac{dv}{dt} = 0$, $u$ and $v$ being constants of motion. Hence the theorem. It is also known as the Jacobi–Poisson theorem. The Poisson theorem is useful in uncovering new constants of motion.

---

## 4.6 Angular momentum

If $l_x, l_y, l_z$ are the components of the angular momentum vector $\vec{l}$, then the Poisson bracket of $l_x$ and $l_y$ may be obtained as

$$\begin{aligned}
\{l_x, l_y\} &= \{yp_z - zp_y, zp_x - xp_z\} \\
&= \{yp_z, zp_x\} - \{yp_z, xp_z\} - \{zp_y, zp_x\} + \{zp_y, xp_z\} \\
&= y\{p_z, z\}p_x + x\{z, p_z\}p_y
\end{aligned}$$

while the other terms vanish. Hence

$$\{l_x, l_y\} = l_z \tag{4.27}$$

Similarly

$$\{l_y, l_z\} = l_x, \quad \{l_z, l_x\} = l_y \tag{4.28}$$

If $l_x$ and $l_y$ are constants of motion then from Poisson's theorem it follows that their Poisson bracket namely $l_z$ is also a constant of motion.

We summarize some additional other results on the Poisson brackets involving $l_x, l_y, l_z$:

$$\begin{aligned}
\{x, l_x\} &= 0 \quad & \{x, l_y\} = z \quad & \{x, l_z\} = -y \\
\{y, l_x\} &= -z \quad & \{y, l_y\} = 0 \quad & \{y, l_z\} = x \\
\{z, l_x\} &= y \quad & \{z, l_y\} = -x \quad & \{z, l_z\} = 0
\end{aligned} \tag{4.29}$$

If $\vec{V} = (V_x, V_y, V_z)$ denotes a constant three-dimensional vector then

$$\begin{aligned} \{x, \vec{V}\cdot\vec{l}\} &= V_y z - V_z y = (\vec{V}\times\vec{r})_x \\ \{y, \vec{V}\cdot\vec{l}\} &= V_z x - V_x z = (\vec{V}\times\vec{r})_y \\ \{z, \vec{V}\cdot\vec{l}\} &= V_x y - V_y x = (\vec{V}\times\vec{r})_z \end{aligned} \tag{4.30}$$

The above results imply that

$$\left\{\vec{r}, \vec{V}\cdot\vec{l}\right\} = \vec{V}\times\vec{r} \tag{4.31}$$

Similarly we can derive

$$\left\{\vec{p}, \vec{V}\cdot\vec{l}\right\} = \vec{V}\times\vec{p} \tag{4.32}$$

The transition from (4.31) to (4.33) involves two compensatory sign changes: changing $\vec{r}$ and $\vec{p}$ produces a change of sign not only in the definition of the Poisson bracket but also in the definition of the angular momentum.

**Example** 4.7

A particle of mass $m$ is acted upon by a constant force $F$. Using the techniques of the Poisson bracket show that $x$ and $p$ are in the forms

$$\begin{aligned} x &= x_0 + \frac{p_0}{m}t + \frac{1}{2}\frac{F}{m}t \\ p &= p_0 + Ft \end{aligned}$$

where $x_0$ and $p_0$ are the initial values of $x$ and $p$.

Let $g$ be a function of the canonical variables $(q, p)$. A Taylor series expansion gives the representation

$$g = g_0 + \left(\frac{dg}{dt}\right)_0 t + \left(\frac{d^2g}{dt^2}\right)_0 \frac{t^2}{2!} + ...$$

where $g_0$ is the initial value of $g$.

Now since

$$\begin{aligned} \frac{dg}{dt} &= \{g, H\} \\ \frac{d^2g}{dt^2} &= \left\{\frac{dg}{dt}, H\right\} = \{\{g, H\}, H\} \end{aligned}$$

and so on, we can represent $g$ as

$$g = g_0 + \{g_0, H\}t + \{\{g_0, H\}, H\}\frac{t^2}{2!} + ...$$

For the given problem since the particle is acted upon by a constant force $F$, we can identify the Hamiltonian as

$$H = \frac{p^2}{2m} - Fx$$

implying for $g = x$ the following results:

$$\begin{aligned} \{x, H\} &= \left\{x, \frac{p^2}{2m} - Fx\right\} = \frac{1}{2m}\{x, p^2\} = \frac{p}{m}\{x, p\} = \frac{p}{m} \\ \{\{x, H\}, H\} &= \left\{\frac{p}{m}, \frac{p^2}{2m} - Fx\right\} = -\frac{F}{m}\{p, x\} = \frac{F}{m} = \text{constant} \end{aligned}$$

So the series for $g = x$ terminates beyond $\{\{x, H\}, H\}$ whence it follows that

$$x = x_0 + \frac{p_0}{m}t + \frac{1}{2}\frac{F}{m}t^2$$

Similarly choosing $g = p$ we find $p = p_0 + Ft$.

**Example 4.8**

Find the position $x$ and momentum $p$ at time $t$ for the motion of a particle given by the Hamiltonian $H = \frac{p^2}{2m} + \frac{1}{2}m\omega^2 x^2$.

We proceed in a similar manner as in the problem above and obtain

$$x = x_0\left(1 - \frac{1}{2!}\omega^2 t^2 + ...\right) + \frac{p_0}{m\omega}\left(\omega t - \frac{1}{3!}\omega^3 t^3 + ...\right) = x_0 \cos\omega t + \frac{p_0}{m\omega}\sin\omega t$$

$$p = -m\omega x_0 \sin\omega t + p_0 \cos\omega t$$

---

## 4.7 Liouville's theorem

Hamilton's canonical equations can be used to deduce Liouville's theorem which points to the preservation of volume in the phase space for a set of points in it as the phase space evolves in time.

We focus on an $n$-particle system described in terms of the generalized coordinates $q_1, q_2, , q_n$ having as their conjugate momenta the components $p_1, p_2, ..., p_n$. These coordinates and momenta constitute a $2n$-dimensional vector in the phase space. Consider the following element of volume $dA$ in such a phase space:

$$dA = dq_1 dq_2 ... d_n dp_1 dp_2 ... dp_n \tag{4.33}$$

As the phase space evolves in time the representative point $(q_i, p_i), i = 1, 2, ..., n$ moves to a new location $(\bar{q_i}, \bar{p_i})$ in time $dt$ as given by

$$\bar{q_i} = q_i + \dot{q_i}dt, \quad \bar{p_i} = p_i + \dot{p_i}dt, \quad i = 1, 2, ..., n \tag{4.34}$$

The differentials $dq_i$ and $dp_i$ change according to

$$d\bar{q_i} = dq_i + \frac{\partial \dot{q_i}}{\partial q_i} dq_i dt, \quad d\bar{p_i} = dp_i + \frac{\partial \dot{p_i}}{\partial p_i} dp_i dt, \quad i = 1, 2, ..., n \tag{4.35}$$

The difference in the volume element can be expressed as

$$dA' = dA + [\sum_{i=1}^{n} (\frac{\partial \dot{q_i}}{\partial q_i} + \frac{\partial \dot{p_i}}{\partial p_i})] dA dt \tag{4.36}$$

Using for $\dot{q_i}$ and $\dot{p_i}$ the Hamilton's equations of motion the terms in the squared brackets in the right-hand side cancel out and we are left with the invariance result $dA = dA'$. These terms actually speak of the vanishing of the divergence of the velocity vector $\vec{V}$ having components $(\dot{q_1}, \dot{q_2}, , \dot{q_n}; \dot{p_1}, \dot{p_2}, ..., \dot{p_n})$, that is, $\nabla.\vec{V} = 0$.

To find the implication of the above result let $\rho$ be the density of the phase fluid over the phase space. The following quantity

$$\rho(q_1, q_2, , q_n; p_1, p_2, ..., p_n) dA$$

can be looked upon as the probability that the system is residing in the infinitesimal volume $dq_1 dq_2 ... dq_{3N} dp_1 dp_2, ... dp_{3N}$.

The continuity equation for $\rho$

$$\frac{\partial \rho}{\partial t} + \sum_{i=1}^{N} (\frac{\partial(\rho \dot{q_i})}{\partial q_i} + \frac{\partial(\rho \dot{p_i})}{\partial p_i}) = 0 \tag{4.37}$$

can be expanded to read

$$\frac{\partial \rho}{\partial t} + \sum_{i=1}^{n} (\frac{\partial \rho}{\partial q_i} \dot{q_i} + \frac{\partial \rho}{\partial p_i} \dot{p_i}) + \rho \sum_{i=1}^{N} (\frac{\partial \dot{q_i}}{\partial q_i} + \frac{\partial \dot{p_i}}{\partial p_i}) = 0 \tag{4.38}$$

Adjusting the second term in the left-hand side in terms of Poisson brackets by using Hamilton's equations of motion which also enable us to do away with the third term we arrive at Liouville's equation

$$\frac{d\rho}{dt} = \frac{\partial \rho}{\partial t} + \{\rho, H\} = 0 \tag{4.39}$$

It shows that $\rho$ is stationary, i.e., the density of the phase fluid is unchanged.

## 4.8 The case of singular Lagrangians

Sometimes we run into a possibility when not all the canonical momenta are independent. Consider a set of $n$ canonical momenta $(p_i, p_2, ..., p_i, ...p_k, ..., p_n)$, where $p_k$ is given by

$$p_k = \frac{\partial L(q_i, \dot{q}_i)}{\partial \dot{q}_k}, \quad i, k = 1, 2, ...n \tag{4.40}$$

$q_i$ and $\dot{q}_i$ being the usual generalized coordinates and velocities.

A Jacobian matrix can be introduced according to

$$[\Delta_{ki}] = \frac{\partial p_k}{\partial \dot{q}_i} = \frac{\partial^2 L}{\partial \dot{q}_i \partial \dot{q}_k} \tag{4.41}$$

keeping in mind that the mapping from the Lagrangian to the Hamiltonian form requires replacement of the velocities $\dot{q}$'s by the canonical momenta $p$'s. For the independence of the $n$ momenta it is of course necessary that $\det(\Delta) \neq 0$. Indeed if we do not have an invertible matrix, then there is no unique solution of the equations of motion. In point of fact, dependence of a few momenta on others will cause the determinant to vanish.

From the Euler–Lagrange equation

$$\frac{d}{dt}\left(\frac{\partial L(q_i, \dot{q}_i)}{\partial \dot{q}_k}\right) - \frac{\partial L(q_i, \dot{q}_i)}{\partial q_k} = 0 \tag{4.42}$$

explicit evaluation of the time derivative gives

$$\frac{\partial^2 L}{\partial q_i \partial \dot{q}_k}\dot{q}_i + \frac{\partial^2 L}{\partial \dot{q}_i \partial \dot{q}_k}\ddot{q}_i - \frac{\partial L}{\partial q_k} = 0 \tag{4.43}$$

which is of the form

$$\Delta_{ik}\ddot{q}_i = \frac{\partial L}{\partial q_k} - \frac{\partial^2 L}{\partial q_i \partial \dot{q}_k}\dot{q}_i \tag{4.44}$$

This reads as a first-order differential equation in the velocities $v_i = \dot{q}_i$

$$\Delta_{ik}\dot{v}_i = \frac{\partial L}{\partial q_k} - \frac{\partial^2 L}{\partial q_i \partial \dot{q}_k}v_i \equiv h_i(q_i, v_i) \tag{4.45}$$

where $h_i$ is a function of coordinates and velocities. Such an equation can be looked upon as the transformed version of the Euler–Lagrangian equation.

Now if the determinant of the Jacobian matrix vanishes there would be existence relationships among the momenta. These could exist as functions

$$\phi_j(p, q, t) = 0, \quad j = 1, 2, ...m \tag{4.46}$$

Such connections speak of *primary constraints* of a Hamiltonian theory. For instance if two momenta $p_1$ and $p_2$ turn out to be equal, resulting from some choice of the Hamiltonian, an obvious choice for $\phi$ would be $\phi \equiv p_1 - p_2$.

Before we turn to a concrete example dealing with such a situation let us have a look at Dirac's procedure of handling the Hamiltonian of singular systems subject to constraints like (4.46). Dirac proposed an extended Hamiltonian of the form

$$H = \bar{H} + \sum_i v_k \phi_k \tag{4.47}$$

where $\bar{H}$ corresponds to the Legendre transformed form

$$\bar{H} = \sum_i p_i \dot{q}_i - L \tag{4.48}$$

and is free from the velocities which are present in the singular part of (4.29).

If we go for a virtual variation, $\delta\bar{H}$ for the right-hand side of Legendre's form above would be

$$\delta\bar{H} = \sum_i \dot{q}_i \delta p_i + \sum_i \delta\dot{q}_i p_i - \sum_i \frac{\partial L}{\partial q_i}\delta q_i - \sum_i \frac{\partial L}{\partial \dot{q}_i}\delta\dot{q}_i = \sum_i (\dot{q}_i \delta p_i - \dot{p}_i \delta q_i) \tag{4.49}$$

with the $\delta\dot{q}_i$ terms canceling out. But being a function of the $q$'s and $p$'s the variation of $\bar{H}$ in itself gives

$$\delta\bar{H} = \sum_i \left(\frac{\partial\bar{H}}{\partial q_i}\delta q_i + \frac{\partial\bar{H}}{\partial p_i}\delta p_i\right) \tag{4.50}$$

Hence on comparison of the two expressions of $\delta\bar{H}$ leads to

$$\sum_{i=1}^{n} \left[\left(\dot{q}_i - \frac{\partial\bar{H}}{\partial p_i}\right)\delta p_i - \left(\dot{p}_i + \frac{\partial\bar{H}}{\partial q_i}\right)\delta q_i\right] = 0 \tag{4.51}$$

On the other hand, variation of (4.46) shows

$$\sum_{i=1}^{n} \left[\frac{\partial\phi_j}{\partial p_i}\delta p_i + \frac{\partial\phi_j}{\partial q_i}\delta q_i\right] = 0, \quad j = 1, 2, ..., m \tag{4.52}$$

The last two equations are consistent if $\dot{q}$ and $\dot{p}$ are given by

$$\dot{q}_i = \frac{\partial\bar{H}}{\partial p_i} + \sum_{i=1}^{n} v_l \frac{\partial\phi_l}{\partial p_i} = \frac{\partial H}{\partial p_i} = \{q_i, H\}, \quad i = 1, 2, ..., n \tag{4.53}$$

$$\dot{p}_i = -\frac{\partial\bar{H}}{\partial q_i} - \sum_{i=1}^{n} v_l \frac{\partial\phi_l}{\partial q_i} = -\frac{\partial H}{\partial q_i} = \{p_i, H\}, \quad i = 1, 2, ..., n \tag{4.54}$$

which one has to remember that $H$, the total Hamiltonian, is given by the extended form (4.47). Note that to build up a viable theory, the primary constraints $\phi_j$ are to be supplemented by the secondary restriction $\dot{\phi}_j = 0$ because like the primary constraints the time derivative of these constraints must vanish at all times.

---

## 4.9 Higher derivative classical systems

Development of higher-derivative classical systems has a long history dating back to an early work of Mikhail Vasilyevich Ostrogradski who extended the Hamilton–Lagrange theory to a generalized solvable form. We give a brief exposition to the procedure of deriving the Lagrangian form of any second-order ordinary differential equation. Basically in the simplest case of a mechanical system with one degree of freedom, the point of enquiry is under what condition a given family of differential equations of second-order, namely,

$$\ddot{q} = F(t, q, \dot{q}) \tag{4.55}$$

acquires a Lagrangian interpretation in the sense that (4.55) can be represented in the manner

$$\frac{\partial L}{\partial q} - \frac{d}{dt}\left(\frac{\partial L}{\partial \dot{q}}\right) = 0 \tag{4.56}$$

where $L$, the Lagrangian, is a function of $q, \dot{q}$ and $t$. When expanded (4.56) reads

$$\frac{\partial L}{\partial q} - \frac{\partial^2 L}{\partial t \partial \dot{q}} - \dot{q}\frac{\partial^2 L}{\partial q \partial \dot{q}} - \ddot{q}\frac{\partial^2 L}{\partial \dot{q}^2} = 0. \tag{4.57}$$

When confronted with (4.45), the above equation takes the form

$$\frac{\partial L}{\partial q} - \frac{\partial^2 L}{\partial t \partial \dot{q}} - \dot{q}\frac{\partial^2 L}{\partial q \partial \dot{q}} - F(t, q, \dot{q})\frac{\partial^2 L}{\partial \dot{q}^2} = 0, \tag{4.58}$$

where $\ddot{q}$ has been replaced by $F$ for consistency. It then follows easily that if we take a partial derivative with respect to $\dot{q}$ and slightly readjust the terms then we arrive at the form

$$\frac{d}{dt}M + \frac{\partial F}{\partial \dot{q}}M = \frac{\partial M}{\partial t} + \frac{\partial}{\partial q}(M\dot{q}) + \frac{\partial}{\partial \dot{q}}(MF) = 0, \tag{4.59}$$

where the quantity $M$ is defined by

$$M(t,q,\dot{q}) = \left(\frac{\partial^2 L}{\partial \dot{q}^2}\right), \tag{4.60}$$

$M$ is referred to as the Jacobi last multiplier ($JLM$).

In terms of $M$, the Lagrangian of a second-order differential equation admits the following structure:

$$L(t,q,\dot{q}) = \int^{\dot{q}} \int^{\dot{q}'} M(t,q,\dot{q}'')d\dot{q}''d\dot{q}' + A(t,q)\dot{q} + B(t,q). \tag{4.61}$$

The simplest choice $M = 1$ shows that the corresponding Lagrangian can be at most a quadratic function of $\ddot{q}$:

$$L = \frac{1}{2}\ddot{q}^2 + A(t,q,\dot{q})\ddot{q} + B(t,q,\dot{q}) \tag{4.62}$$

---

## 4.10 The Pais–Uhlenbeck oscillator

The Pais–Uhlenbeck oscillator is a suitable special example obeying fourth-order dynamical equation of motion

$$\ddddot{q} + \alpha\ddot{q} + \beta q = 0 \tag{4.63}$$

where $\alpha = (\omega_1^2 + \omega_2^2)$ and $\beta = \omega_1^2\omega_2^2$ are real and positive parameters. Let us replace it by an equivalent system of two coupled differential equations that follow from the usual first-order Euler–Lagrange equation.

Defining a new set of coordinates

$$Q_1 := q, \quad \dot{Q}_1 = \dot{q}; \quad \text{and} \quad Q_2 := \ddot{q} \tag{4.64}$$

one can apply the usual Euler–Lagrange equation

$$\frac{d}{dt}\left(\frac{\partial L}{\partial \dot{Q}_i}\right) - \frac{\partial L}{\partial Q_i} = 0, \quad i = 1, 2 \tag{4.65}$$

along with the Lagrangian

$$L = -\frac{1}{2}Q_2^2 - \frac{1}{2}\alpha\dot{Q}_1^2 - \dot{Q}_1\dot{Q}_2 + \frac{1}{2}\beta Q_1^2 \tag{4.66}$$

to get to a coupled second-order system

$$\alpha\ddot{Q}_1 + \ddot{Q}_2 + \beta Q_1 = 0, \quad \ddot{Q}_1 = Q_2 \tag{4.67}$$

Eliminating the variable $Q_2$ gives the desired equation for PUO.

The Lagrangian allows us to define the conjugate momenta in the usual way

$$P_i = \frac{\partial L}{\partial \dot{Q}_i}, \quad i = 1, 2 \tag{4.68}$$

With the Hamiltonian

$$H = P_i \dot{Q}_i - L \tag{4.69}$$

and canonical momenta given by

$$P_1 = -\alpha \dot{Q}_1 - \dot{Q}_2, \quad P_2 = -\dot{Q}_1 \tag{4.70}$$

a Hamiltonian for the PUO system can be extracted using $L$ given by (4.66)

$$H = -P_1 P_2 + \frac{1}{2}\alpha P_2^2 + \frac{1}{2}Q_2^2 - \frac{1}{2}\beta Q_1^2 \tag{4.71}$$

From the standard definition of the Poisson bracket

$$\dot{Q}_i = \{Q_i, H\}, \quad \dot{P}_i = \{P_i, H\}, \quad i = 1, 2 \tag{4.72}$$

the first-order Hamilton equations are given by

$$\dot{Q}_1 = -P_2, \quad \dot{Q}_2 = -P_1 + \alpha P_2; \quad \dot{P}_1 = \beta Q_1, \quad \dot{P}_2 = -Q_2 \tag{4.73}$$

---

## 4.11 Summary

In this chapter we gave a brief description of the Hamiltonian formalism in classical mechanics that forms the basis for working in a phase space. We derived Hamiltonian canonical equations of motion in terms of the generalized coordinates and generalized or canonical momenta. Unlike Lagrange's equations of motion which hold in the configuration space, Hamilton's equations are a set of two coupled partial differential equations representing the motion of the system in the phase space. We introduced Poisson brackets and reviewed their main properties including recasting Hamiltonian equations in terms of the Poisson bracket. We also discussed Liouville's theorem. Further we touched upon the subject of singular Lagrangians and gave an exposition to higher derivative classical systems.

# Exercises

(1) If the Hamiltonian of a dynamical system is given by $H = p_1 q_1 - p_2 q_2 - aq_1^2 + bq_2^2$ where $a, b$ are constants, show that $\frac{p_2 - bq_2}{q_1}$ =constant.

(2) If $H = qp^2 - pq + bp$, find $q$ and $p$ as functions of $t$. Here $b$ is a constant.

(3) The Lagrangian for the motion of a particle in a rotating frame is

$$L = \frac{1}{2} m \dot{\vec{r}}^2 + \frac{1}{2} m\omega^2 (x_1^2 + x_2^2) + m\omega(x_1 \dot{x_2} - x_2 \dot{x_1})$$

Find the Hamiltonian.

(4) For a linear transformation from the canonical variables $(q, p)$ to the pair $(Q, P)$ given by $Q = a_{11} q + a_{12} p, P = a_{21} q + a_{22} p$, find the conditions on the numbers $a_{ij}, i, j = 1, 2$ so that $\{q.p\} = 1$ and $p\dot{q} = P\dot{Q} + \frac{dF}{dt}$ where $F$ is a function of $q$ and $p$.

(5) Consider a particle of mass $m$ attached to one end of a string placed in a vertical position. The tension of the spring is subject to Hooke's law. The other end of the string is attached to a fixed point. Suppose that the spring constant is $k$ and its natural length is $l$. Write down the Lagrangian and Hamiltonian.

(6) The point of suspension of a simple pendulum of length $l$ and mass $m$ is constrained to move on a parabola given by the equation $\xi = ax^2, \quad a > 0$ in the vertical plane. Derive the Lagrangian and Hamiltonian governing the motion of the pendulum and its point of suspension.

(7) A Hamiltonian of a mechanical system with one degree of freedom has the form

$$H = \frac{p^2}{2a} - \mu qpe^{-at} + \frac{a\mu}{2} q^2 e^{-at}(a + \mu e^{-at}) + \frac{k}{2} q^2$$

where $a, \mu, k$ are constants, $q$ is the generalized coordinate and $p$ is the generalized momentum. Find the Lagrangian and an equivalent Lagrangian which is not explicitly time-dependent.

(8) Set up Hamiltonian equations for the Lagrangian

$$L(q, \dot{q}, t) = \frac{m}{2}(\dot{q}^2 \sin^2 \omega t + q\dot{q}\omega \sin 2\omega t + q^2 \omega^2)$$

(9) If $\vec{X}$ is defined as $\vec{X} = c_1 \vec{r} + c_2 \vec{p} + c_3(\vec{r} \times \vec{p})$ where $c_1, c_2, c_3$ are arbitrary constants, then show for an arbitrary vector $\vec{V}$ that the Poisson bracket $\{\vec{X}, \vec{V}.\vec{l}\} = \vec{V} \times \vec{X}$, where $\vec{l}$ is the angular momentum vector.

(10) Consider a particle of mass $m$ and charge $e$ moving in a magnetic field $\vec{A}$ and described by the Lagrangian

$$L = \frac{1}{2}m\dot{r}^2 + \frac{e}{c}\dot{r} \cdot \vec{A}$$

where $c$ is a constant. Derive the following Poisson brackets:

$$\{v_x.v_y\} = \frac{e}{cm^2}A_z, \quad \{v_y.v_z\} = \frac{e}{cm^2}A_x, \quad \{v_z.v_x\} = \frac{e}{cm^2}A_y,$$

where $\vec{v}$ is given by $\vec{v} = \frac{\vec{p}-\frac{e}{c}\vec{A}}{m}$.

# Chapter 5

## Dynamical systems: An overview



A dynamical system is generally concerned with an evolutionary problem related to some physical, chemical or biological system in which the values of the underlying variables are specified in an ordered way. A dynamical system can describe a continuous evolution process addressing input and output variables through a differential equation (linear or nonlinear) or a discrete one that is controlled by difference equations or even deal with hybrid schemes which integrate models of continuous and discrete dynamics. In this chapter we will focus only on the continuous class of dynamical systems. One can consider such systems analytically or subject them to numerical simulations when tracking their behavior goes out of hand chiefly because of the uncontrollable nonlinear influence. A dynamical system is deterministic or random depending on when being subject to a specific evolution rule the system is taken uniquely from a given definite state to a subsequent one or fails to be deterministic in which case a stochastic differential equation can guide it. Mathematically, nonlinear equations do not normally have solutions which would superpose making the systems they represent rather complicated and difficult to analyze. In fact, most nonlinear systems are not analytically tractable and it is impossible to predict for an arbitrarily long time the dynamical properties of even weakly perturbative series. In a nonlinear theory multiple equilibrium points can occur. One also frequently encounters such terms as the limit cycles, high-period orbits and the phenomenon of chaos. For instance, a limit cycle typically refers to an isolated closed trajectory in the sense that when one looks at its neighboring trajectories they are not closed: either they tend to diverge away from the limit cycle or spiral toward it. On the other hand, a chaotic system is hypersensitive to initial conditions. In fact, a very slight difference in the initial conditions can cause the trajectories to follow different lines of evolution.

---

## 5.1 Basic notions and preliminaries

In general, a dynamical system refers to an initial value problem of the kind

$$\frac{d\vec{X}}{dt} = \vec{f}(\vec{X}, t, u) \tag{5.1}$$

where $\vec{X}(t)$ stands for an $m$-dimensional vector and $\vec{f}$ is a vector field which can also depend on $\vec{X}$ and $t$ apart from a set of auxiliary variables collectively denoted as $u$. More specifically $\vec{X}$ and $\vec{f}$ are represented by

$$\vec{X}(t) \equiv \begin{pmatrix} x_1(t) \\ x_2(t) \\ \ldots \\ \ldots \\ x_m(t) \end{pmatrix}, \quad \vec{f} \equiv \begin{pmatrix} f_1 \\ f_2 \\ \ldots \\ \ldots \\ f_m \end{pmatrix} \tag{5.2}$$

Prey-predator systems, bacterial growth and radioactive decay of atoms are some simple examples of a dynamical system.

In principle $f$ can depend upon several other parameters like the coupling constant, amplitude, etc. but these have been omitted from giving an explicit display. We assume that we have a deterministic system at hand and the set of initial conditions $(x_1(0), x_2(0), ...x_m(0))$ serves as judicious inputs to determine uniquely the solution we are after. Normally these initial conditions are always subject to uncontrollable fluctuations in the measurements because no matter how good our intention is we can never make an infinitely precise measurement. Thus, measurements are always subject to uncertainties so that errors are bound to creep in. The question then becomes relevant as to how sensitive are the solutions to small errors accompanying the initial conditions. We take the evolution rule to be guided by a map from the underlying phase space to itself. We parameterize such a map by the time parameter $t$ and refer to it by the function $\varphi_t(x)$.

Any differential equation is characterized by its order. Thus, (5.1) represents a system with order $m$. The simplest one corresponds to $m = 1$ which is a straightforwardly integrable case:

$$x = x(0) + \int_0^t f(s, x(s), u)ds \tag{5.3}$$

for some initial condition $x(0)$. The general solution for all $t$ corresponding to the set of initial conditions $(x_1(0), x_2(0), ...x_m(0))$ is called a complete flow or simply a flow and the set of points along a flow is called the trajectory or an orbit. A flow at time $t$ is the mapping $\varphi_t(x)$. We note that $\varphi_0(x_0) = x_0$ and that the composition rule for the flow, namely, $\varphi_t \circ \varphi_s = \varphi_{t+s}$ holds. Here the composition means $\varphi_t \circ \varphi_s = \varphi_t(\varphi_s(x))$.

If $t$ is not explicitly present in (5.1) the system is classified as an autonomous system. Autonomous systems are thus stationary and in the absence of the set of auxiliary variables $u$, $\vec{X}$ is given by

$$\frac{d\vec{X}}{dt} = \vec{f}(x_1, x_2, ...x_m) \tag{5.4}$$

We say that the vector field $\vec{f}$ is subject to the Lipschitz condition on a set $S$ if for any pair of coordinates $(\vec{X}^{(i)}, \vec{X}^{(j)}) \in S \subset R^m$ there exists a constant $K$ such that

$$|\vec{f}(\vec{X}^{(i)}) - \vec{f}(\vec{X}^{(j)})| < K|\vec{X}^{(i)} - \vec{X}^{(j)}| \tag{5.5}$$

for all $K > 0$. If the vector field $\vec{f}$ is subject to the Lipschitz condition in $S$, then there exists a unique solution for the differential equation for a given set of initial conditions.

The set of Hamilton's equations of motion defined in terms of the Hamiltonian $H = H(x_1, x_2, ..., x_m; p_1, p_2, ..., p_m)$ but not consisting of $t$ explicitly, namely,

$$\begin{aligned} \dot{x}_i &\equiv \frac{dx_i}{dt} = \frac{\partial H}{\partial p_i} \\ \dot{p}_i &\equiv \frac{dp_i}{dt} = -\frac{\partial H}{\partial x_i} \end{aligned} \tag{5.6}$$

$(i = 1, 2, ..., m)$ is an example of a conservative autonomous dynamical system with $\vec{f}$ standing for the column matrix

$$\vec{f} \equiv \begin{bmatrix} \frac{\partial H}{\partial p_1} \\ \frac{\partial H}{\partial p_2} \\ \cdots \\ \cdots \\ \cdots \\ \frac{\partial H}{\partial p_m} \\ \frac{-\partial H}{\partial x_1} \\ \frac{-\partial H}{\partial x_2} \\ \cdots \\ \cdots \\ \cdots \\ \frac{-\partial H}{\partial x_m} \end{bmatrix} \tag{5.7}$$

More compactly, we can reexpress (5.5) in the manner

$$\frac{d\vec{X}}{dt} = A\vec{f}, \quad A = \begin{pmatrix} 0 & +I_m \\ -I_m & 0 \end{pmatrix} \tag{5.8}$$

where $\vec{X}$ is a column matrix like the one given in (5.2) whose first $m$ entries are the coordinates $(x_1, x_2, ...x_m)$ and the remaining $m$ entries are the momenta $(p_1, p_2, ...p_m)$. $A$ is written in a block form with the object $I_m$ representing for the $(m \times m)$ identity matrix.

State variables are those that offer a complete description of the state of a dynamical system while the state space corresponds to the set of all the possible values of the state variables. The dimension of the dynamical system is given by the number of state variables. Phase space normally refers to the state space which is continuous and finite-dimensional. The phase plane consists of trajectories obtained from the solutions of (5.1). For instance, taking $m = 1$, a phase plane is guided by the pair $(x, \frac{dx}{dt})$ while for $m = 2$, the phase plane corresponds to $(x_1, x_2)$ or simply $(x, y)$ where $x$ and $y$ are a pair of Cartesian coordinates. An ensemble of trajectories for a given system in the phase plane or a phase space is called a phase portrait (see Figure 5.1). An isocline refers to a curve in a phase plane or a phase space on which the trajectories have a fixed gradient. In general, the set of initial conditions $(x_1(0), x_2(0), ...x_m(0))$ dictates how a dynamical system evolves ultimately toward the final state (which may be a point or an area or a curve depending on the manifold the dynamical system operates in) in the phase space or state space. Such a destination is called an attractor.

An equilibrium point or a stationary point is a solution resulting from the vanishing of the right-hand side of (5.1). Thus, if $\bar{\vec{x}}$ is an equilibrium or a fixed point we have

$$\vec{f}(t, \bar{\vec{x}}, u) = 0 \tag{5.9}$$

An equilibrium point or a stationary point is also referred to as a fixed point or a critical point. The equilibrium may or may not be stable. If say, the origin $(0, 0)$ represents a state of stable equilibrium toward which the set of all the trajectories is driven at, then the state of stable equilibrium of the origin is termed the point attractor or a sink. If the opposite is the case, we have a repeller or a source at the origin.

We now turn to the following examples.

**Example 5.1**

Consider a simple dynamical system

$$\dot{x} = f(x), \quad f(x) = \sin x \tag{5.10}$$

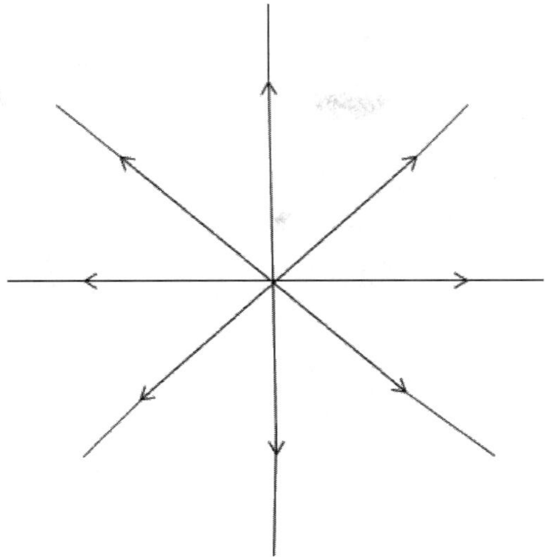

**FIGURE 5.1**: Phase portrait.

The fixed points are at $x = n\pi$, $n = 0, \pm 1, \pm 2, ....$ The integration can be performed easily starting with an initial value of $x_0 = \frac{\pi}{4}$ to arrive at the following form for $t$:

$$t = \ln |\frac{1+\sqrt{2}}{\csc x + \cot x}| \tag{5.11}$$

While it is not immediately clear about the large-$t$ behavior of $x(t)$, it is instructive to do a geometrical analysis to have an idea of the stable and unstable nature of the fixed point.

**Example 5.2**

Consider the damped pendulum problem where we modify the pendulum equation (1.68) by including the effect of a frictional term which is proportional to the velocity of the motion

$$\ddot{\theta} + \lambda\dot{\theta} + \frac{g}{l}\sin\theta = 0 \tag{5.12}$$

where $\lambda$ is a damping constant. We observe that (5.12) can be put in the framework of a dynamical system by casting it as a pair of coupled first-order differential equations by setting $x = \theta$ and $y = \dot{\theta}$:

$$\dot{x} = y, \quad \dot{y} = -\frac{g}{l}\sin x - \lambda y \tag{5.13}$$

In this way we can convert almost any higher-order differential equation into a system of first order ones by introducing suitable intermediate variables. In this particular problem, it is the time-derivative of $x$, namely, the variable $y$, which serves the purpose. Of course such a procedure increases the dimensions of the system. Note that in interpreting (5.12) in terms of (5.13) we have enhanced the dimension by two.

The fixed points of (5.13) are identified by setting the right-hand side of each equation to zero and are readily obtained as the set of points $(n\pi, 0)$, $n = 0, \pm 1, \pm 2, \ldots$. Note that while $(0,0)$ is a stationary solution which conforms to the usual stable equilibrium position of the pendulum hanging straight down, the solution $(\pm\pi, 0)$ is stationary but unstable since they point to the straight up position of the pendulum.

We therefore see that the fixed points could be stable or unstable. Our next point of inquiry will be to look into certain stability aspects of some simple systems of classical mechanics.

---

## 5.2 Simple examples from classical mechanics

The one-dimensional Newton's second law of motion in the presence of a potential $V(x)$ is described by a pair of dynamical equations

$$\dot{x} = \frac{p}{m}, \quad \dot{p} = -\frac{dV}{dx} \tag{5.14}$$

where $x$ is the position and $p$ is the momentum of the particle. An extremum of the potential occurs when $\dot{p} = 0$.

If $(x_0, p_0)$ is an equilibrium position of the system then, as we saw in (1.41), Taylor series expansion about $x = x_0$ gives for small values of $(x - x_0)$ an approximation of $V(x)$

$$V(x) \approx V(x_0) + (x - x_0)V'(x_0) + \frac{1}{2}(x - x_0)^2 V''(x_0) \tag{5.15}$$

Such a representation of $V(x)$ when plugged in the expression for the momentum $p$ as furnished in (1.91) gives

$$(p - p_0)^2 = 2m[E - E_0 - \frac{1}{2}(x - x_0)^2 V''(x_0)] \tag{5.16}$$

where we assume that $E$ has the value $E_0$ at the point $(x_0, p_0)$ and use the criterion that at an extremum point $V'(x_0) = 0$.

Two possibilities immediately present themselves from the above expression. When the potential has a minimum at $(x = x_0)$, then $V''(x_0) > 0$, the system being subjected to an attractive force, and the locus of the representative point describes elliptic trajectories with center at $(x_0, y_0)$

$$\frac{(y - y_0)^2}{b^2} = 1 - \frac{(x - x_0)^2}{a^2} \tag{5.17}$$

where $a^2 = \frac{2(E-E_0)}{V''(x_0)}$ and $b^2 = 2m(E-E_0)$ are, respectively, the squares of the semi-major and semi-minor axis of the ellipse. We call the point $(x_0, p_0)$ in the phase plane an elliptic singularity for $E > E_0$ which is in a state of stable equilibrium.

On the other hand, when $V''(x_0) < 0$, which corresponds to the possibility of a repulsive force, the representative point traces out hyperbolic trajectories as given by

$$\frac{(y-y_0)^2}{b^2} = -1 + \frac{(x-x_0)^2}{a^2} \tag{5.18}$$

with respect to the center $(x_0, y_0)$. Note that $\pm\frac{b}{a}$ are the slopes of the asymptotes that pass through the center and $a$ is the fixed distance of the vertices from the center. The point $(x_0, p_0)$ in the phase plane is called a hyperbolic singularity or a saddle. The motion in the neighborhood of $(x_0, y_0)$ is in a state of unstable equilibrium.

Take the case of the harmonic oscillator when the particle executes a simple harmonic motion being attracted to a given fixed point by a constant force $F = -kx, \quad k > 0$, which is proportional to the distance from the point. The harmonic oscillator is a conservative system and as was already noted in Chapter 1, it admits of a general trigonometric solution $x(t) = A\cos(\omega_0 t + \lambda)$, where $\omega_0 = \sqrt{\frac{k}{m}}$, implying that $p(t) = -Am\omega_0 \sin(\omega_0 t + \lambda)$. With the potential for such a conservative system being $V(x) = \frac{1}{2}m\omega_0^2 x^2$, the energy $E$ turns out to be $E = \frac{p^2}{2m} + \frac{1}{2}m\omega_0^2 x^2 = \frac{mA^2\omega_0^2}{2}$. Squaring and adding the expression for $x$ and $p$ we readily find that the representative point traces out paths of ellipse in the phase plane with semi-major and semi-minor axis of lengths $(\frac{1}{\omega_0}\sqrt{\frac{2E}{m}}, \sqrt{2Em})$. We identify the point $(0, 0)$ as the point of stable equilibrium.

When the force is repulsive, i.e., $F = -kx, \quad k < 0$, a different picture emerges. Here the trigonometric solutions are to be replaced by their hyperbolic ones and we run into the possibility of two different combinations of solutions, namely, $x(t) = \pm A\cosh(\omega_0 t)$ and $p(t) = \pm Am\omega_0 \sinh(\omega_0 t)$ or the pair $x(t) = \pm A\sinh(\omega_0 t)$ and $p(t) = \pm Am\omega_0 \cosh(\omega_0 t)$. In the first case, the energy is $E = \frac{p^2}{2m} - \frac{1}{2}m\omega_0^2 x^2 = -\frac{mA^2\omega_0^2}{2}$ which is negative while in the second case, the energy is $E = +\frac{mA^2\omega_0^2}{2}$ which is positive. For the negative energy solution we have at $t = 0$, $x = \pm A$ and $p = 0$ which means that the particle during the course of its motion from $x = \pm\infty$ passes through the points $x = \pm A$ and subsequently moves away in the future. For the positive energy solution, however, the particle takes up the position $x = 0$ at $t = 0$ with a finite value of $p = \pm Am\omega_0$ on the center of repulsion in the phase plane. It is clear that the asymptotes, which intersect at the point of unstable equilibrium $(0, 0)$, branch out the regions of positive and negative energy solutions.

What happens in the case of the simple pendulum problem? The equation of motion we wrote down in (1.68), namely,

$$\ddot{\theta} + \frac{g}{l}\sin\theta = 0 \tag{5.19}$$

has the accompanying Hamiltonian

$$H = \frac{1}{2}\dot{\theta}^2 - \frac{g}{l}\cos\theta \tag{5.20}$$

From the above form of $H$ we can identify different types of regions for the motion. In the interval $(-\frac{g}{l}, \frac{g}{l})$, the motion is finite and oscillatory in character and corresponds to curves for the oscillatory pendulum motion swinging back and forth about its vertical position. However, while $H < -\frac{g}{l}$ does not allow for any physical motion, $\dot{\theta}$ being negative in this case, the other extreme corresponding to $H > -\frac{g}{l}$ points to an unbounded motion. $H = \frac{g}{l}$ corresponds to the separatrix which separates the phase space, formed by the pair of variables $(\theta, \dot{\theta})$, into two distinct regions. It passes through the point $(\pi, 0)$ in the phase space.

To determine the equation of the separatrix we note that on it the Hamiltonian assumes the form

$$\frac{g}{l} = \frac{1}{2}\dot{\theta}^2 - \frac{g}{l}\cos\theta \tag{5.21}$$

which gives for $\dot{\theta}$

$$\dot{\theta} = \pm 2\sqrt{\frac{g}{l}}\cos\frac{\theta}{2} \tag{5.22}$$

The above equation can be easily integrated to yield

$$t = \sqrt{\frac{l}{g}}\ln\tan\frac{1}{4}(\theta + \pi) \tag{5.23}$$

and represents the equation of the separatrix.

Writing (5.23) in the form $e^{\sqrt{\frac{g}{l}}t} = \tan(\frac{\theta}{4} + \frac{\pi}{4})$ with a similar one for its inverse, and then adding the two it is straightforward to obtain the relationship

$$\cos\frac{\theta}{2} = \text{sech}\sqrt{\frac{g}{l}}t \tag{5.24}$$

Thus from (5.22) we arrive at the form of the velocity profile

$$\dot{\theta} = \pm 2\sqrt{\frac{g}{l}}\text{sech}\sqrt{\frac{g}{l}}t \tag{5.25}$$

It represents a bell-shaped curve resembling a soliton-like solitary wave. The latter decays asymptotically with time while its width of the curve goes inversely as the factor $\sqrt{\frac{g}{l}}$.

## 5.3 Analysis of a linear system

Consider a system of $n$ linear equations as given by

$$\begin{aligned}\dot{x_1} &= a_{11}x_1 + a_{12}x_2 + ... + a_{1n}x_n \\ \dot{x_2} &= a_{21}x_1 + a_{22}x_2 + ... + a_{2n}x_n \\ & \quad .... \\ \dot{x_n} &= a_{n1}x_1 + a_{n2}x_2 + ... + a_{nn}x_n \end{aligned} \tag{5.26}$$

with constant coefficients $a_{ij}$, $i, j = 1, 2, ...n$. We assume $\det(a_{ij}) \neq 0$ of the coefficient matrix $A_{ij}$. Looking for solutions in the exponential forms

$$x_1 = \alpha_1 e^{\beta t}, x_2 = \alpha_2 e^{\beta t}, ...x_n = \alpha_n e^{\beta t} \tag{5.27}$$

with constant $\alpha_k$'s, $k = 1.2....n$ and $\beta$, the set of equations (5.13) is transformed to

$$\begin{aligned}(a_{11} - \beta)\alpha_1 + a_{12}\alpha_2 + ... + a_{1n}\alpha_n &= 0 \\ a_{21}\alpha_1 + (a_{22} - \beta)\alpha_2 + ... + a_{2n}\alpha_n &= 0 \\ & .... \\ a_{n1}\alpha_1 + a_{n2}\alpha_2 + ... + (a_{nn} - \beta)\alpha_n &= 0 \end{aligned} \tag{5.28}$$

From the theory of linear algebraic equations, we know that (5.15) will admit of nontrivial solutions in $\alpha_k$'s provided the determinant of the coefficient matrix vanishes:

$$\det \begin{pmatrix} a_{11} - \beta & a_{12} & ... & a_{1n} \\ a_{21} & a_{22} - \beta & ... & a_{2n} \\ ... & ... & ... & ... \\ a_{n1} & a_{n2} & ... & a_{nn} - \beta \end{pmatrix} = 0 \tag{5.29}$$

The above equation is referred to as the characteristic equation for $\beta$. Its roots are the eigenvalues of the matrix $A$.

Let us discuss the simplest case corresponding to $n = 2$ for the matrix $A$. In the Cartesian $(x, y)$ plane we address the two equations

$$\begin{aligned}\dot{x} &= a_{11}x + a_{12}y \\ \dot{y} &= a_{21}x + a_{22}y \end{aligned} \tag{5.30}$$

for $-\infty < t < +\infty$, having the underlying determinant $\Delta \equiv a_{11}a_{22} - a_{12}a_{21}$.

The characteristic equation for such a reduced case turns out to be a quadratic in $\beta$, namely,

$$\beta^2 - \beta(a_{11} + a_{22}) + \Delta = 0 \tag{5.31}$$

If $\beta_1$ and $\beta_2$ are the two roots of this equation, then these are given by

$$\beta_1, \beta_2 = \frac{1}{2}[TrA \pm \sqrt{(TrA)^2 - 4\Delta}] \tag{5.32}$$

In other words,

$$\beta^2 - \beta(a_{11} + a_{22}) + \Delta \equiv (\beta - \beta_1)(\beta - \beta_2) \tag{5.33}$$

with

$$TrA = (a_{11} + a_{22}) = (\beta_1 + \beta_2), \quad \Delta = a_{11}a_{22} - a_{12}a_{21} = \beta_1\beta_2 \tag{5.34}$$

We call $D$ to be the discriminant

$$D = (TrA)^2 - 4\Delta \tag{5.35}$$

and it is easily found that $D = (\beta_1 - \beta_2)^2$.

Several possibilities present themselves depending upon the sign of the discriminant $D$ and the sign of the determinant $\Delta$. If $\Delta > 0$, then the following obvious three cases turn up:

(a) $\Delta = \beta_1\beta_2 > 0$, $D > 0$

Here $\beta_1$ and $\beta_2$ are real and distinct eigenvalues keeping the same sign. If $TrA < 0$ and the exponentials in (5.14) act as damping factors (sink), then the system is stable for asymptotic $t$. The solution is referred to as the stable node (see Figure 5.2). If $TrA > 0$ and the exponential factors in (5.14) grow with time (source), the opposite situation is encountered: the system is unstable

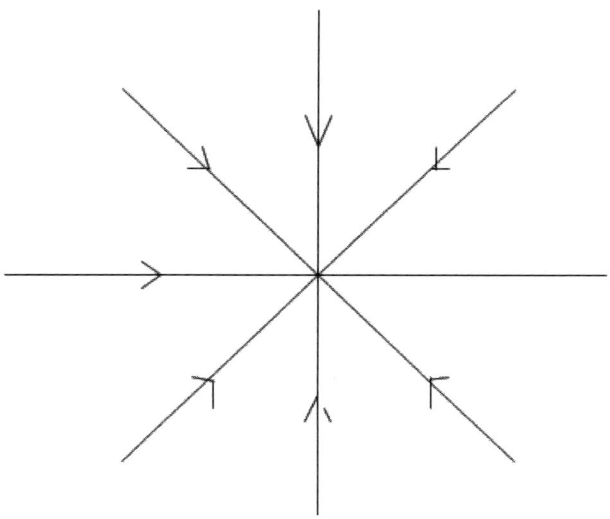

**FIGURE 5.2**: Stable node.

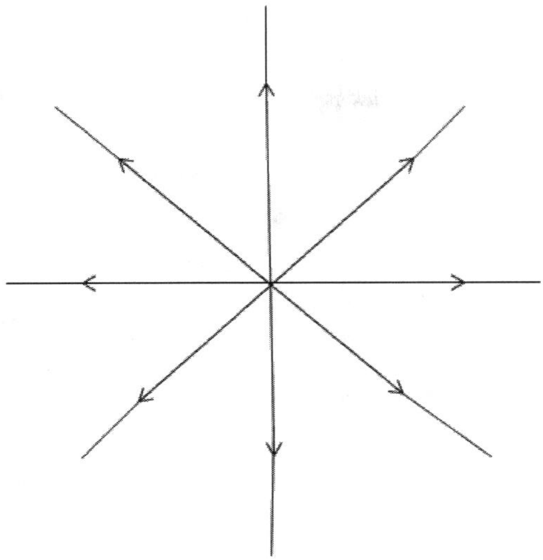

**FIGURE 5.3**: Unstable node.

for asymptotic $t$. The solution is referred to as the unstable node (see Figure 5.3). If $TrA$ =0, then $\beta_1$ =- $\beta_2$ implying $\Delta = -{\beta_1}^2 < 0$ contrary to the chosen sign of $\Delta$ and hence is not a possible mode.

(b) $\Delta = \beta_1\beta_2{>}0$, $D{<}0$

Here complex eigenvalues are signalled. Let these be

$$\beta_1 = \gamma + i\delta, \quad \beta_2 = \gamma - i\delta \tag{5.36}$$

where the imaginary part $\delta \neq 0$. If $TrA$ <0, then it implies $\gamma < 0$ and we have an asymptotically stable case. The solution is referred to as the stable focus or a stable spiral point (see Figure 5.4). The contrary behavior is noticed for $TrA$ >0 when $\gamma > 0$. Such a case corresponds to an unstable focus or an unstable spiral point (see Figure 5.5). If $TrA$ =0, for which $\gamma = 0$, the solutions are guided only by a phase term $e^{\pm i\delta t}$ suggesting only closed orbits around the fixed point. The solution is stable in character and referred to as the neutrally stable center rather than the asymptotically stable fixed point (see Figure 5.6).

(c) $\Delta = \beta_1\beta_2{>}0$, $D{=}0$

This is the case of equal roots $\beta_1 = \beta_2$. We carry out the analysis as follows: If there are two linearly independent eigenvectors, then $(TrA) < 0$ corresponds to $\beta_1 = \beta_2 < 0$ and we have asymptotic stability which is that of a nodal point and referred to as the stable star node. In this connection it needs to be noted that if any of the coefficients of our basic equation (5.17) are fluctuated a little, then this may cause a slight displacement of the equal roots to a pair of conjugate complex roots or a pair of real roots. Thus, the position of a stable star node is intermediate between a nodal point or a focal

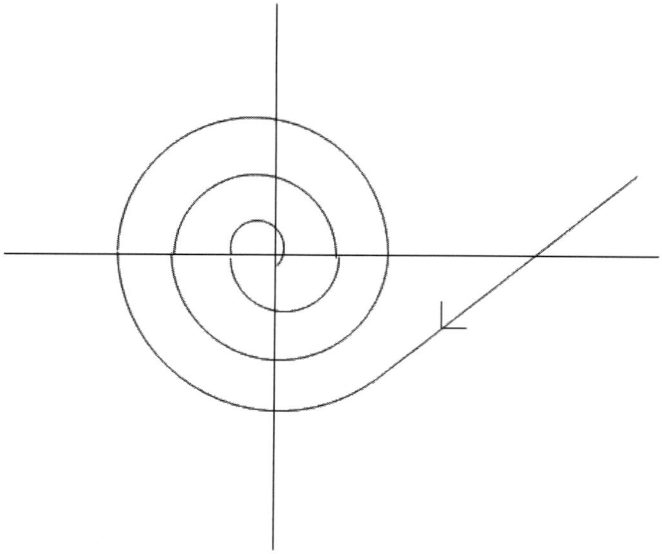

**FIGURE 5.4**: Stable focus or a stable spiral point.

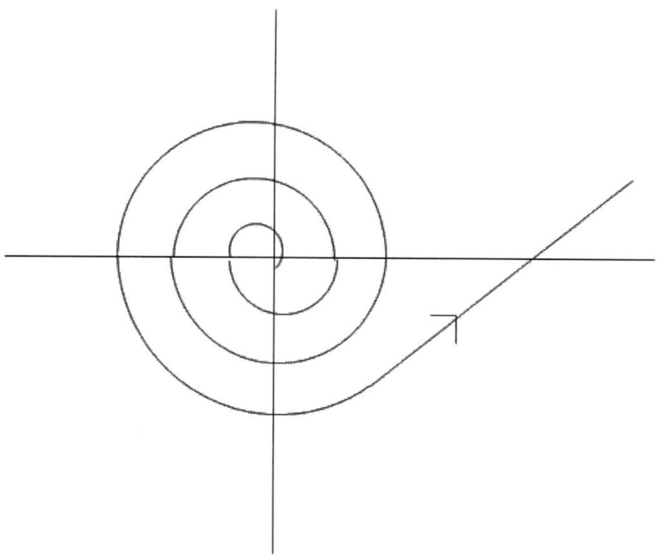

**FIGURE 5.5**: Unstable focus or unstable spiral point.

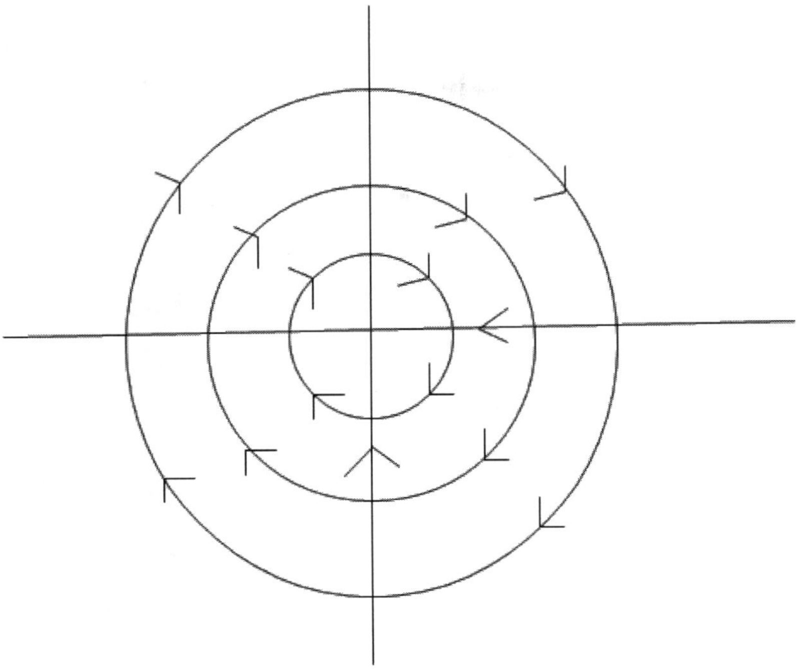

**FIGURE 5.6**: Neutrally stable center.

point. If however $(TrA) > 0$, then the opposite behavior holds and we have an unstable star node. The case $(TrA) = 0$ points to $\beta_1 =- \beta_2$ which cannot hold unless both the eigenvalues vanish pointing to $\Delta =0$ which is contrary to the assumption $\Delta > 0$. If there is only one independent eigenvector, then the role of stable star node and unstable star node are replaced by stable degenerate node and unstable degenerate node, respectively.

Next we turn to $\Delta = \beta_1\beta_2 < 0$. Here the eigenvalues are of opposite signs showing that while one of the exponentials in (5.14) decays, the other grows. We thus have a combination of trajectories one of which asymptotically (with time) converges to the fixed point as represented by the negative eigenvalue and the other diverges (hyperbolic in nature) as represented by the positive eigenvalue. With the presence of one diverging term due to the effect of the positive eigenvalue, the overall stability cannot be assured and so the fixed point is an unstable saddle point (see Figure 5.7).

Finally we discuss the case when $\Delta = \beta_1\beta_2 = 0$. It follows that at least one eigenvalue is zero and we get a line of fixed points. To clarify such an issue we consider first the case of $TrA <0$ with $\beta_1 = 0$ and $\beta_2 < 0$. It gives a line of neutrally stable (non-isolated) fixed points. On the other hand, for the case $TrA >0$ the opposite feature holds and we have a line of unstable

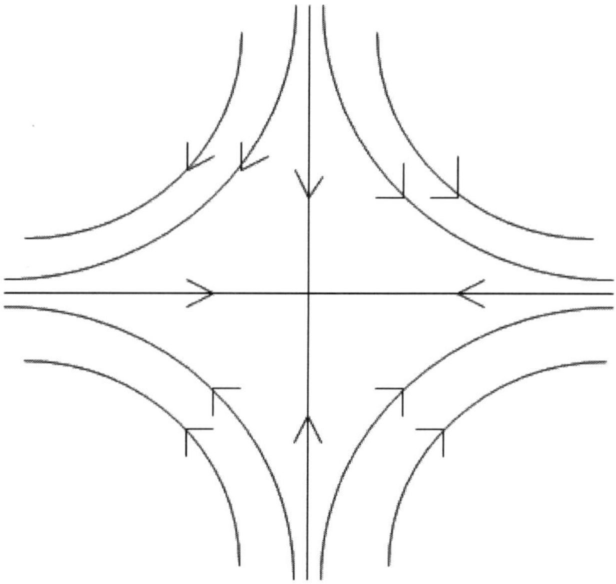

**FIGURE 5.7**: Unstable saddle point.

(non-isolated) fixed points. For $TrA = 0$, the two eigenvalues simultaneously acquire a zero value. If there are two linearly independent eigenvectors, then we have a plane of neutrally stable (non-isolated) fixed points while for only one independent eigenvector, the plane of neutrally stable (non-isolated) fixed points is replaced by a line of unstable (non-isolated) fixed points.

Some comments about the nature of the general solutions. For two distinct real eigenvalues, the general solution can be expressed as

$$\vec{x} = c_1\vec{\alpha_1}e^{\beta_1 t} + c_2\vec{\alpha_2}e^{\beta_2 t} \tag{5.37}$$

where $c_1$ and $c_2$ are constants and $\vec{\alpha_1}$ and $\vec{\alpha_2}$ are two independent eigenvectors.

When there are repeated real eigenvalues with two linearly independent eigenvectors the general solution can be expressed as

$$\vec{x} = (c_1\vec{\alpha_1} + c_2\vec{\alpha_2})e^{\beta t} \tag{5.38}$$

where $c_1$ and $c_2$ are constants, $\beta_1 = \beta_2 = \beta$, and $\vec{\alpha_1}$ and $\vec{\alpha_2}$ are two independent eigenvectors. For the case when there is only one independent eigenvector $\vec{\alpha}$, the general solution reads

$$x = (c_1\vec{\alpha} + c_2\vec{\alpha}t)e^{\beta t} \tag{5.39}$$

where $c_1$ and $c_2$ are constants.

For the conjugate complex eigenvalues the general solution is expressed as

$$x(t) = e^{\gamma t}(c_1 \cos\delta t + c_2 \sin\delta t)\vec{p} + e^{\gamma t}(c_1{}' \cos\delta t + c_2{}' \sin\delta t)\vec{q} \tag{5.40}$$

where $c_1$ and $c_2$ are constants, $\vec{p}$ and $\vec{q}$ are two independent eigenvectors and $c_1{}'$ and $c_2{}'$ are another set of constants related to $c_1$ and $c_2$.

We illustrate the above notions by means of the following and subsequent examples.

**Example 5.3**

Let us consider the damped oscillator problem as considered earlier in (1.43)

$$\ddot{x} + \omega_0^2 x + \lambda\omega_0\dot{x} = 0 \tag{5.41}$$

where $\lambda$ is proportional to the damping constant incorporating the effects of the friction.

We can interpret the above second-order equation as a combination of two first order equations

$$\dot{x} = y, \quad \dot{y} = -\omega_0^2 x - \lambda\omega_0 y \tag{5.42}$$

By equating to zero the above expressions for $\dot{x}$ and $\dot{y}$, the fixed point is clearly seen to be at the origin $(0, 0)$.

The accompanying characteristic equation is evaluated from the determinant

$$\det\begin{pmatrix} -\beta & 1 \\ -\omega_0{}^2 & -\lambda\omega_0 - \beta \end{pmatrix} = \beta^2 + \lambda\omega_0\beta + \omega_0^2 \equiv 0 \tag{5.43}$$

It gives for the eigenvalues

$$\beta_1, \beta_2 = -\frac{\lambda\omega_0}{2} \pm \frac{\omega_0}{2}\sqrt{\lambda^2 - 4} \tag{5.44}$$

First of all, if $\lambda = 0$, then the system reduces to the simple harmonic oscillator depicting a periodic motion. The eigenvalues are given by $\pm i\omega_0$. We therefore have a center at the origin. Next with $\lambda > 2$ (overdamping), the fixed point $(0, 0)$ is a stable node. It includes also the case of critical damping for $\lambda = 2$. However, in the range $0 < \lambda < 2$ (underdamping), the fixed point $(0, 0)$ is a stable focus. Finally, corresponding to $\lambda < 0$, we have an unstable focus and unstable node for the fixed point $(0, 0)$ corresponding to the respective cases of $\lambda^2 < 4$ and $\lambda^2 \geq 4$.

**Example 5.4**

Consider a linear system as given by

$$\dot{x} = x - y, \quad \dot{y} = 7x + 5y - 3 \tag{5.45}$$

The fixed point is obtained by setting the derivatives $\dot{x}$ and $\dot{y}$ equal to zero. We therefore get a unique solution $(\frac{1}{4}, \frac{1}{4})$. For a linear system the matrix $A$ is a constant. The eigenvalues are obtained by solving the characteristic equation which for this system is easily found to be $\beta^2 - 6\beta + 12 = 0$. The roots are in conjugate complex pairs namely $3 \pm i\sqrt{3}$. It has a positive real part and hence can be identified as an unstable focus or a spiral source. To find the eigenvectors we solve for the matrix equation

$$\begin{pmatrix} 1-\beta & -1 \\ 7 & 5-\beta \end{pmatrix} \begin{pmatrix} \chi \\ \varsigma \end{pmatrix} = \begin{pmatrix} 0 \\ 0 \end{pmatrix} \tag{5.46}$$

for $\beta_{1,2} = 3 \pm i\sqrt{3}$. For the positive sign we obtain the following link between $\chi$ and $\varsigma$:

$$\varsigma = -(2 + i\sqrt{3}) \tag{5.47}$$

implying that the accompanying two eigenvectors are

$$\alpha^{(1)} = \begin{pmatrix} 1 \\ -2 - i\sqrt{3} \end{pmatrix} \tag{5.48}$$

and replacing $i$ by -$i$, the second one

$$\alpha^{(2)} \begin{pmatrix} 1 \\ -2 + i\sqrt{3} \end{pmatrix} \tag{5.49}$$

Hence the solution provided by the first eigenvector $\alpha^{(1)}$ is

$$= e^{(3+i\sqrt{3})t} \begin{pmatrix} 1 \\ -2 - i\sqrt{3} \end{pmatrix} \tag{5.50}$$

which can also be put as

$$= e^{3t} \begin{pmatrix} \cos\sqrt{3}t + i\sin\sqrt{3}t \\ (2\cos\sqrt{3}t - \sqrt{3}\sin\sqrt{3}t) + i(2\sin\sqrt{3}t + \sqrt{3}\cos\sqrt{3}t) \end{pmatrix} \tag{5.51}$$

where we have used Euler's formula $e^{i\theta} = \cos\theta + i\sin\theta$. Splitting the real and imaginary components this is of the form $\phi_1 + \phi_2$ where

$$\phi_1 = e^{3t} \begin{pmatrix} \cos\sqrt{3}t \\ (2\cos\sqrt{3}t - \sqrt{3}\sin\sqrt{3}t \end{pmatrix}, \quad \phi_2 = e^{3t} \begin{pmatrix} \sin\sqrt{3}t \\ 2\sin\sqrt{3}t + \sqrt{3}\cos\sqrt{3}t \end{pmatrix} \tag{5.52}$$

For the second eigenvector $\alpha^{(2)}$, we would be similarly led to the form $\phi_1 - \phi_2$. Of course for a linear system the sum and difference of two solutions are also solutions and so we conclude that both $\phi_1$ and $\phi_2$ are solutions too. These are real solutions and linearly independent and hence the general solution is obtained as

$$x(t) = c_1 e^{3t} \begin{pmatrix} \cos\sqrt{3}t \\ (2\cos\sqrt{3}t - \sqrt{3}\sin\sqrt{3}t \end{pmatrix} + \quad c_2 e^{3t} \begin{pmatrix} \sin\sqrt{3}t \\ 2\sin\sqrt{3}t + \sqrt{3}\cos\sqrt{3}t \end{pmatrix} \tag{5.53}$$

where $c_1$ and $c_2$ are constants which can be determined from some given initial conditions.

---

## 5.4 Nonlinear systems: Process of linearization

Let a pair of coupled differential equations be given by

$$\dot{x} = ax + by + f(x, y), \quad \dot{y} = cx + dy + g(x, y) \tag{5.54}$$

that has as its linear portion the equations

$$\dot{x} = ax + by, \quad \dot{v} = cx + dy \tag{5.55}$$

In (5.54) and (5.55), $a, b, c$ and $d$ are some real constants. The linearization is carried out about a certain relevant point. Choosing it to be a fixed point $(x_0, y_0)$, we assume that $f$ and $g$ are continuously, differentiable functions about it. We also assume that the ratios $\frac{f(x,y)}{r}$ and $\frac{g(x,y)}{r}$ where $r = \sqrt{(x - x_0)^2 + (y - y_0)^2}$ are well behaved at the origin in the sense that both of them approach 0 as $r \to 0$.

Let us consider a two-dimensional nonlinear system

$$\frac{d\vec{X}}{dt} = A\vec{f} \tag{5.56}$$

where $\vec{X}$ and $\vec{f}$ are given by the column matrices

$$\vec{X}(t) \equiv \begin{pmatrix} x(t) \\ y(t) \end{pmatrix}, \quad \vec{f} \equiv \begin{pmatrix} f_1(x, y) \\ f_2(x, y) \end{pmatrix} \tag{5.57}$$

Suppose $(x_0, y_0)$ is a fixed point about which we carry out a Taylor expansion for $f_1(x, y)$ and $f_2(x, y)$ as follows:

$$f_1(x, y) = f_1(x_0, y_0) + (x - x_0)(\frac{\partial f_1}{\partial x})_0 + (y - y_0)(\frac{\partial f_1}{\partial y})_0 + R_1(x, y) \tag{5.58}$$

$$f_2(x, y) = f_2(x_0, y_0) + (x - x_0)(\frac{\partial f_2}{\partial x})_0 + (y - y_0)(\frac{\partial f_2}{\partial y})_0 + R_2(x, y) \tag{5.59}$$

where $R_1(x,y)$ and $R_2(x,y)$ are the remainder terms controlled by the conditions

$$\lim_{r\to 0}[\frac{R_1(x,y)}{r}] = 0, \quad \lim_{r\to 0}[\frac{R_2(x,y)}{r}] = 0 \tag{5.60}$$

where $r = \sqrt{(x-x_0)^2 + (y-y_0)^2}$.

Since $(x_0, y_0)$ is a fixed point, it should satisfy the stationary condition and hence $f_1(x_0, y_0)$ and $f_2(x_0, y_0)$ (being just constants) have to vanish. Transforming to new variables $X$ and $Y$ by performing a shift about the fixed point namely

$$X = x - x_0, \quad Y = y - y_0 \tag{5.61}$$

it readily follows that $\dot{X}$ and $\dot{Y}$ are given by

$$\dot{X} = X(\frac{\partial f_1}{\partial x})_0 + Y(\frac{\partial f_1}{\partial y})_0 + R_1(X + x_0, Y + y_0) \tag{5.62}$$

$$\dot{Y} = X(\frac{\partial f_2}{\partial x})_0 + Y(\frac{\partial f_2}{\partial y})_0 + R_2(X + x_0, Y + y_0) \tag{5.63}$$

On identifying

$$a = (\frac{\partial f_1}{\partial x})_0, b = (\frac{\partial f_1}{\partial y})_0, c = (\frac{\partial f_2}{\partial x})_0, d = (\frac{\partial f_2}{\partial y})_0 \tag{5.64}$$

and neglecting the remainder terms $R_1$ and $R_2$ we achieve a linear approximation of the nonlinear system (5.62) and (5.63) but in terms of $X$ and $Y$ variables. We indeed obtain

$$\frac{d}{dt}\begin{pmatrix} X \\ Y \end{pmatrix} = M \begin{pmatrix} X \\ Y \end{pmatrix} \tag{5.65}$$

where the two-dimensional matrix $M$ given by

$$M = \begin{pmatrix} (\frac{\partial f_1}{\partial x})_0 & (\frac{\partial f_1}{\partial y})_0 \\ (\frac{\partial f_2}{\partial x})_0 & (\frac{\partial f_2}{\partial y})_0 \end{pmatrix} \tag{5.66}$$

is called the Jacobian matrix. The main task is to evaluate the Jacobian matrix at the fixed point and go for the linearization. Let us take some simple situations in the following examples.

**Example 5.5**

Suppose a dynamical system is described by the equations

$$\dot{x} = 3x - x^2 - xy, \quad \dot{y} = 2y - xy - 2y^2 \tag{5.67}$$

By putting $\dot{x} = 0$ and $\dot{y} = 0$ we determine the fixed points to be $(0,0)$ and $(4,-1)$. Note that here, according to our notations, $f_1 = 3x - x^2 - xy$ and

$f_2 = 2y - xy - 2y^2$. About the point $(0,0)$, we get $\left(\frac{\partial f_1}{\partial x}\right)_0 = 3, \left(\frac{\partial f_1}{\partial y}\right)_0 = 0, \left(\frac{\partial f_2}{\partial x}\right)_0 = 0$ and $\left(\frac{\partial f_2}{\partial y}\right)_0 = 2$. Hence the matrix $M$ is diagonal and given by

$$M = \begin{pmatrix} 3 & 0 \\ 0 & 2 \end{pmatrix} \tag{5.68}$$

It has positive eigenvalues 3 and 2. Hence the fixed point is an unstable node.

What about the other fixed point $(4, -1)$? Since at this point the values of the partial derivatives are $\left(\frac{\partial f_1}{\partial x}\right)_0 = -4, \left(\frac{\partial f_1}{\partial y}\right)_0 = -4, \left(\frac{\partial f_2}{\partial x}\right)_0 = 1$ and $\left(\frac{\partial f_2}{\partial y}\right)_0 = -2$, the corresponding matrix $M$ is given by

$$M = \begin{pmatrix} -4 & -4 \\ 1 & -2 \end{pmatrix} \tag{5.69}$$

By solving the characteristic equation (5.17), we get a complex pair of roots for $\beta$ with a negative real part as given by $-3 \pm i\sqrt{3}$. The fixed point is therefore a stable focus.

**Example 5.6**

Let us consider the system

$$\dot{x} = 2x - xy - 2x^2, \quad \dot{y} = 2x + xy \tag{5.70}$$

Here $f_1 = 2x - 2x^2 - xy$ and $f_2 = 2y + xy$. The fixed points are easily determined to be $(0,0)$ and $(-2, 6)$ by putting $\dot{x} = 0$ and $\dot{y} = 0$. For the fixed point $(0,0)$, the eigenvalue 2 is repeated and the diagonal matrix turns out to be

$$M = \begin{pmatrix} 2 & 0 \\ 0 & 2 \end{pmatrix} \tag{5.71}$$

The general solution is of the type

$$c_1 e^{2t} \begin{pmatrix} 1 \\ 0 \end{pmatrix} + c_2 e^{2t} \begin{pmatrix} 0 \\ 1 \end{pmatrix} \tag{5.72}$$

where $c_1$ and $c_2$ are constants. Here there are two linearly independent eigenvectors with a positive trace. Hence we have an unstable star mode.

For the other fixed point $(-2, 6)$, the partial derivatives are $\left(\frac{\partial f_1}{\partial x}\right)_{(2,-6)} = 4, \left(\frac{\partial f_1}{\partial y}\right)_{(2,-6)} = 2, \left(\frac{\partial f_2}{\partial x}\right)_{(2,-6)} = 6$ and $\left(\frac{\partial f_2}{\partial y}\right)_{2,-6)} = 0$. The corresponding matrix $M$ is given by

$$M = \begin{pmatrix} 4 & 2 \\ 6 & 0 \end{pmatrix} \tag{5.73}$$

and the characteristic equation reads $\beta^2 - 4\beta - 12 = 0$. Its roots are 6 and $-2$. These are of opposite signs and we conclude that the fixed point is an unstable saddle point.

**Example 5.7**

For the system

$$\dot{x} = x - 18xy + 10x^2 + 7, \quad \dot{y} = -11y + 2xy + 9 \tag{5.74}$$

we easily see by inspection that by putting $\dot{x} = 0$ and $\dot{y} = 0$ a fixed point is located at the point $(1, 1)$. To analyze its character we evaluate the partial derivatives and find $(\frac{\partial f_1}{\partial x})_{(1,1)} = 3, (\frac{\partial f_1}{\partial y})_{(1,1)} = -18, (\frac{\partial f_2}{\partial x})_{(1,1)} = 2$ and $(\frac{\partial f_2}{\partial y})_{1,1)} = -9$. Hence the corresponding matrix $M$ is given by

$$M = \begin{pmatrix} 3 & -18 \\ 2 & -9 \end{pmatrix} \tag{5.75}$$

The charateristic equation turns out to be $(\beta + 3)^2 = 0$ implying that the eigenvalue $-3$ is repeated. The eigenvectors are obtained by solving the matrix equation

$$\begin{pmatrix} 3-\beta & -18 \\ 2 & -9-\beta \end{pmatrix} \begin{pmatrix} \xi \\ \eta \end{pmatrix} = \begin{pmatrix} 0 \\ 0 \end{pmatrix} \tag{5.76}$$

at the point $\beta = -3$. We obtain as a result the relationship $\xi = 3\eta$ implying only one eigenvector is involved. Since the trace is negative we have a stable degenerate node for this fixed point.

Suppose now we slightly perturb the elements of the matrix $M$ in the manner

$$M = \begin{pmatrix} 3 & -18+\epsilon \\ 2 & -9 \end{pmatrix} \tag{5.77}$$

As a result of the presence of a small nonzero quantity $\epsilon$ the characteristic equation changes to $\beta^2 + 6\beta + 9 - 2\epsilon = 0$ showing that the equal eigenvalues $-3$ branch out to different values $-3 \pm \sqrt{2\epsilon}$. For $\epsilon > 0$, we get distinct negative eigenvalues implying a stable node while for $\epsilon < 0$, complex eigenvalues appear with a negative real part signalling a stable spiral.

**Example 5.8**

Our heart is a rather complicated nonlinear system. Its periodic oscillatory behavior in normal situations must be stable as our everyday experience shows. A somewhat realistic heartbeat model was proposed in 1973 by E.C. Zeeman which we now discuss briefly. As we all know the systolic (or the period of chamber contraction) and diastolic (or the period of chamber relaxation) are the two phases of the heart. The pulsating of the heart is due to the electrochemical gradient which taps each individual fiber. During systolic contraction the heart squeezes signalling consumption of energy while during relaxation energy is needed to overcome the bond (or the threshold) taking place during contraction that moves out the calcium ions out of the systole.

In this way the regulation of the pace of the heart goes on in a cycle with the heart contracting from a diastole to a systole and then rapidly back again to the original diastolic state which is a state of stable equilibrium (fixed point).

To make a qualitative study of heart functioning Zeeman considered the following scheme in terms of the equations

$$\epsilon\dot{x} = x - x^3 - y, \quad \dot{y} = x - x_0$$

where $x$ stands for the length of a cardiac muscle fiber but has a small positive coefficient $\epsilon$ prefixed due to the fast eigenvalue of the system, $y$ is a signature of electrochemical activity and $x_0$ is a constant corresponding to a typical length in the muscle fiber in the diastolic state with $x_0 > \frac{1}{\sqrt{3}}$.

By putting $\dot{x} = 0$ and $\dot{y} = 0$ we determine the equilibrium point at $(x_0, y_0)$ where $y_0 = x_0 - x_0^3$. To examine its stability behavior, we need to carry out linearization and study the Jacobian matrix of this system.

Using the notations already employed earlier, we find here $f_1 = x - x^3 - y$ and $f_2 = x - x_0$. The partial derivatives turn out to be $(\frac{\partial f_1}{\partial x})_{(x_0,y_0)} = 1 - 3x_0^2$, $(\frac{\partial f_1}{\partial y})_{(x_0,y_0)} = -1$, $(\frac{\partial f_2}{\partial x})_{(x_0,y_0)} = 1$ and $(\frac{\partial f_2}{\partial y})_{(x_0,y_0)} = 0$. Hence the corresponding Jacobi matrix $M$ is given by

$$M = \begin{pmatrix} 1 - 3x_0^2 & -1 \\ 1 & 0 \end{pmatrix} \tag{5.78}$$

It yields the characteristic equation

$$\beta^2 - (1 - 3x_0^2)\beta + 1 = 0$$

from which the two eigenvalues can be solved in the forms

$$\beta_{1,2} = \frac{1}{2}[(1 - 3x_0^2) \pm \sqrt{(1 + 3x_0^2)(x_0^2 - 1)}]$$

Interestingly both the eigenvalues are negative for the possibilities $x_0 > \frac{1}{\sqrt{3}}$ or $x_0 < -\frac{1}{\sqrt{3}}$. In each case the equilibrium point $(x_0, y_0)$ is stable. This will be clear from the remarks made below.

A graphical illustration of the model in the $yx$-plane is shown in Figure 5.8. The heartbeat cycle has two equilibrium states, namely, the diastole and systole. The main point to note is that as the chemical control variable $y$ increases, there is a gradual contraction in the muscle fiber until a threshold position is reached when there is a rapid contraction and a systolic equilibrium state is reached. After this there is a rapid relaxation of the muscle fibers until the heart is returned to the original diastolic state and the cycle is completed.

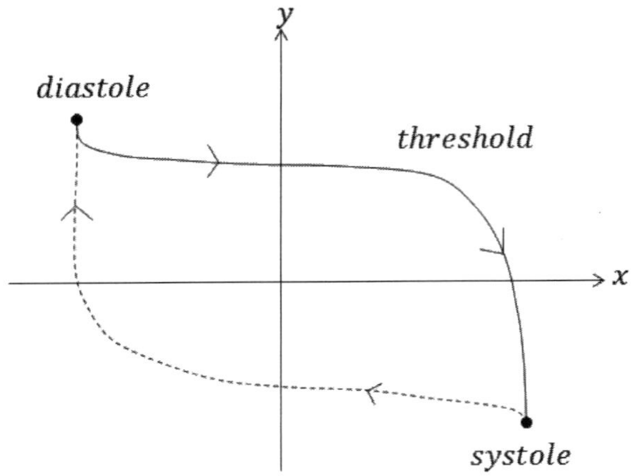


**FIGURE 5.8**: The heartbeat model.

## 5.5 Lotka–Volterra model

Study of interacting populations in a given environment is of great interest in problems of mathematical ecology and constitutes an active area of research. The interactions affect the growth rates of the organism in the ecosystem and so for a balance between various types of species an appropriate predator-prey relationship needs to be maintained. Among other things, the following aspects are crucial to our understanding of the qualitative nature of the system. Specifically one needs to know (i) how the interactions among various species manifest themselves, (ii) how the species coexist in a given environment, (iii) how nature strikes a balance between the two and (iv) how a prey escapes being killed by a predator. It is natural that when the preys are too many then the population of the predators increases but in the reverse case when the population of the predators are too numerous then the population of the preys shows signs of decreasing. On the other hand, while it is evident that without preys the predators will reveal a negative growth rate, a limited number of predators implies that the preys will tend to survive over a longer period of time and ultimately reach a stage when an exponential rise in their population will be noticed. In a practical scenario repeated increase and decrease in the population sizes of preys and predators signals a cyclic behavior of the survival of the two species.

The Lotka–Volterra model is one of the simplest schemes of population growth in a closed ecosystem (i.e., derecognizing any migration into or out of the system) which aims at studying the inter-behavior of only two species, the predator and the prey. It may not be out of context to mention here that the spirit of the Lotka–Volterra model was contained in the 1920 paper by Alfred Lotka, an American biologist and actuary, who proposed a predator-prey model to understand the population variation of the interacting fish. Such a study was of utmost relevance then since one already had some knowledge on the percentage of the total catch of selachians (such as sharks, rays, dogfish and skates) which were the predators of the fish (such as garfish, eel and sardines) giving some qualitative idea on the predator-prey interactions. The data collected by the Italian biologist Umberto d'Ancona accounted for the period between 1914 and 1923 and showed a rather substantial increase of the predators during the time of the First World War. Although people initially thought that this could be due to less fishing due to the turmoil of the war, it was also realized that the food fish also revealed marked changes simultaneously. Umberto d'Ancona's father-in-law was Vito Volterra who also worked on the same model as Lotka's around the same time. In 1925 he came up with some explanations of the data collected by Umberto by giving an interpretation of the cause-effect interactions between the selachians (predators) and the food fish (preys) and sought to explore, in general, the cyclic relationship between the predator-prey species. Since then it has been customary to refer to the two species predator-prey model as the Lotka–Volterra model.

In the Lotka–Volterra model, a pair of coupled evolution equations is proposed involving two unknown variables—one for the prey ($x$) and another for the predator ($y$). The degree of interaction is assumed to be proportional to the product $xy$. In this way the following two equations are set up:

$$\dot{x} = ax - gxy, \quad \dot{y} = -by + gxy \tag{5.79}$$

where the parameters $a, b$ and $g$ are all $> 0$. Here the parameter $a$ signifies the natural growth rate of the prey in the absence of any predator, the parameter $b$ is responsible for the natural decay rate of predators if there is no prey to hunt for, and the parameter $g$ stands for the interaction coefficient. It is clear that if the predator is absent, i.e., $y = 0$, the prey will survive and show an exponential growth. However, if there is no prey, i.e., $x = 0$, a negative growth rate is signalled for the predators with the possibility that they may even go extinct. The interaction term in (5.21) addresses the question of both the preys being eaten up by their predators and the effect on the growing size of the latter. In principle $g$ could be different for the two equations but for simplicity we have taken them to be equal. Note that by hunting the rate of increase of prey is reduced (and so $a$ is decreased) while the rate of decrease of predator is increased (and so $b$ is increased) but the interaction coefficient $g$ is not supposed to be affected.

Lotka–Volterra equations can also be transformed to a one-parameter form by effecting the following transfer of variables:

$$s = \frac{gx}{d}, \quad r = \frac{gy}{a}, \quad \tau = at \tag{5.80}$$

We thus obtain

$$\frac{ds}{d\tau} = s(1-r), \quad \frac{dr}{d\tau} = \lambda r(s-1) \tag{5.81}$$

showing $\lambda$ to be the only control parameter of the system.

Further, by defining the variables

$$p = \log s, \quad q = \log r \tag{5.82}$$

we can easily identify the Hamiltonian for (5.26) to be

$$H = \lambda(e^p - p) + (e^q - q) \tag{5.83}$$

Indeed, it can be readily checked from the Hamiltonian equations $\frac{dq}{d\tau} = \frac{\partial H}{\partial p}$ and $\frac{dp}{d\tau} = -\frac{\partial H}{\partial q}$ that relations (5.26) hold.

Turning to (5.21) a conservation law can be worked out by noticing that the two equations yield on division

$$\frac{dy}{dx} = \frac{-by + gxy}{ax - gxy} \tag{5.84}$$

which yields as the first integral the quantity

$$F(x,y) = \log[|x|^b |y|^a e^{-g(x+y)}] = C \tag{5.85}$$

where $C$ is a constant. $F(x,y) = C$ represents the conservation law.

It is interesting to look for the nature of the fixed points of the Lotka–Volterra system. From Equation (5.21) the fixed points are obtained from their stationary versions, namely,

$$ax - gxy = 0, \quad -by + gxy = 0 \tag{5.86}$$

The fixed points are identified to be located at

$$(0,0) \quad \text{and} \quad (\frac{a}{g}, \frac{b}{g}) \tag{5.87}$$

Note that $\frac{dx}{dt} = 0$ gives the horizontal isoclines ($x = 0$ or $y = \frac{a}{g}$) while $\frac{dy}{dt} = 0$ gives the vertical isoclines ($y = 0$ or $x = \frac{b}{g}$).

Now the Jacobian matrix for such a system reads

$$M = \begin{pmatrix} a - gy & -gx \\ gy & -b + gx \end{pmatrix} \tag{5.88}$$

At $(0,0)$, $M$ has the form

$$M = \begin{pmatrix} a & 0 \\ 0 & -b \end{pmatrix} \tag{5.89}$$

showing it to be a diagonal one with obvious eigenvalues $a$ and $-b$. Their differing signs indicate that the fixed point $(0,0)$ is an unstable saddle point.

On the other hand, for the second fixed point $(\frac{a}{g}, \frac{b}{g})$, $M$ is given by

$$M = \begin{pmatrix} 0 & -b \\ a & 0 \end{pmatrix} \tag{5.90}$$

It is an off-diagonal matrix with imaginary eigenvalues given by $\pm i\sqrt{ab}$. Hence the fixed point is a stable center.

## 5.6 Stability of solutions: Lyapunov function

Consider a class of solutions $x(t)$ of the dynamical system as given by Equation (5.3). A fixed equilibrium point $x(t_0)$ at time $t = 0$ is said to be Lyapounov stable if for all $\epsilon > 0$, there exists a $\delta(\epsilon, t_0) > 0$ such that, if around the equilibrium point $|x(0) - x(t_0)| < \delta$ holds true at time $t = 0$, then $|x(t) - x(t_0)| < \epsilon$ also holds true for all times $t \geq 0$. More explicitly, if we think of a neighborhood $K$ around the equilibrium point $x(t_0)$ then such an equlibrium is termed Lyapunov stable if we can find a $K_0 \subset K$ such that for any $x \in K_0$, the flow $\varphi_t(x) \in K$ for all times $t \geq 0$. All periodic orbits display stability of the Lyapunov type. Violation of the Lyapunov condition even for at least one solution renders the equilibrium unstable. Note that the stability is uniform if we can choose a $\delta$ independent of $t_0$, i.e., $\delta(\epsilon)$. Apart from being not only Lyapounov stable, if around the equilibrium point, $|x(0) - x(t_0)| < \delta$ holds true for $\delta > 0$, then the equilibrium is said be asymptotically stable if $\lim_{t\to\infty} |x(t) - x(t_0)| = 0$ is fulfilled. Asymptotic stability has relevance for those dynamical systems if when perturbed slightly around an equilibrium point the system will have the tendency to return to it.

Let us extend the concept of the Hamiltonian function $H(x, y)$ for an autonomous system which is controlled by the pair of Hamilton equations

$$\frac{dx}{dt} = \frac{\partial H}{\partial p}, \quad \frac{dp}{dt} = -\frac{\partial H}{\partial x} \tag{5.91}$$

to an arbitrary continuously differentiable function $U(x, y)$ mimicking the role of the Hamiltonian. In other words, $U$ is subject to similar styled equations as those given above

$$\frac{dx}{dt} = \frac{\partial U}{\partial y}, \quad \frac{dy}{dt} = -\frac{\partial U}{\partial x} \tag{5.92}$$

where the partial derivatives are assumed to exist and are continuous. Then it is evident that $U$ also satisfies being a constant of motion:

$$\frac{dU}{dt} = \frac{\partial U}{\partial x}\frac{dx}{dt} + \frac{\partial U}{\partial y}\frac{dy}{dt} = 0 \tag{5.93}$$

by substituting the above relations for $\frac{dx}{dt}$ and $\frac{dy}{dt}$. At an equilibrium fixed point both the partial derivatives of $U$ vanish. Denoting $\vec{\Phi} \equiv (\frac{\partial U}{\partial y}, -\frac{\partial U}{\partial x})$, it also turns out that $\text{div}\Phi = 0$. Further it also holds that $\frac{dy}{dx}$ is given by

$$\frac{dy}{dx} = -\frac{\frac{\partial U}{\partial x}}{\frac{\partial U}{\partial y}} \tag{5.94}$$

When the denominator $\frac{\partial U}{\partial y}$ vanishes the solution curves terminate at a point.

**Example 5.9**

Consider

$$U - -\frac{1}{2}(x^2 + y^2) \tag{5.95}$$

then evaluating the partial derivatives $\frac{\partial U}{\partial x}$ and $\frac{\partial U}{\partial y}$, we find $\frac{dx}{dt} = -y$ and $\frac{dy}{dt} = x$. Thus the fixed point is $(0,0)$. Now since $\frac{dy}{dx} = -\frac{x}{y}$, when integrated it gives the solution curve as circles $x^2 + y^2 = c^2$ where $c$ is a constant, for $(x,y) \neq (0,0)$.

We now inquire into the possibility of replacing the partial derivatives $\frac{\partial U}{\partial x}$ and $\frac{\partial U}{\partial y}$ by some arbitrary functions say, $u(x,y)$ and $v(x,y)$. Then $\frac{dx}{dt}$ and $\frac{dy}{dt}$ would read

$$\frac{dx}{dt} = u(x,y), \quad \frac{dy}{dt} = v(x,y) \tag{5.96}$$

Hence any function of $x$ and $y$, say, $W(x,y)$, which depends upon the variable $t$ implicitly due to $x$ and $y$ being functions of $t$, will have its time derivative go as

$$\frac{dW}{dt} = \frac{\partial W}{\partial x}\frac{dx}{dt} + \frac{\partial W}{\partial y}\frac{dy}{dt} = u(x,y)\frac{\partial W}{\partial x} + v(x,y)\frac{\partial W}{\partial y} \tag{5.97}$$

where the partial derivatives of $W$ are assumed to exist and are continuous. The above equation can be expressed in the form

$$\frac{dW(\vec{x})}{dt} = \nabla W(\vec{x}) \cdot \vec{\xi} \tag{5.98}$$

where $\vec{x} = (x,y)$ and $\vec{\xi} = (u,v)$. Note that $W(\vec{x})$ is called a Lyapunov function, and it will be our endeavor to seek for such a function for the understanding of the stability of a system. It must however be pointed out that for many nonlinear systems constructing an appropriate Lyapunov function proves very difficult. Nonetheless, there are situations when a Lyapunov function exists in a simple way.

Toward this end we impose upon $W(\vec{x})$ the criterion of being a positive definite function while restricting its time derivative to be negative semi-definite: $\nabla W(\vec{x}) \cdot \vec{\xi} \leq 0$. We then refer to $W(\vec{x})$ as a *weak* Lyapunov function. In the case when the semi-definiteness can be replaced by a strict $\nabla W(\vec{x}) \cdot \vec{\xi} < 0$, $W(\vec{x})$ is referred to as a *strong* Lyapunov function. Let $W_0$ be an equilibrium point of a flow $\varphi_t$. The Lyapunov stability theorem asserts that if $W(\vec{x})$ is a weak Lyapunov function in some region $K$ around the equilibrium point $W_0$, then $W_0$ is (Lyapunov) stable. On the other hand, asymptotic stability is guaranteed if $W(\vec{x})$ turns out to be a strong Lyapunov function.

For a justification of the above statements we proceed as follows. For concreteness, we choose, without any loss of generality, the equilibrium point $W_0$ to correspond to $x = 0$. We surround the point by a small region, say, $K_\epsilon$ of radius $\epsilon$ such that $K_\epsilon \subset K$. The region $K_\epsilon$ is compact and a minimum $\mu$ of $W$ exists in it (i.e., in the region defined by $|x| = \epsilon$) which is positive because $W$ is positive. Since $W$ decreases as $x$ goes to zero, it is clear that we can find a $\delta < \epsilon$ such that the value of $W$ will drop below $\mu$ for $x \in K_\delta$. Now from (5.98), since $\frac{dW}{dt} \leq 0$ for a weak Lyapunov function, $W$ is nonincreasing along orbits $\varphi_t$ and we can infer that $W(\varphi_t(x)) < \mu$ corresponding to $x \in K_\delta$. With $\delta < \epsilon$, $W < \mu$ on $|x| = \epsilon$ as well pointing to $\varphi_t(x) \in K_\delta$. We thus arrive at the conclusion that the equilibrium point $x = 0$ is stable in the sense of Lyapunov.

To prove the asymptotic stability we make use of the Bolzano–Weierstrass theorem which states that every bounded infinite set of real numbers has at least one limit point. Consider a sequence of time points $t_1, t_2, ..., t_n, ....$ Then the compactness of $K_\epsilon$ implies that the corresponding sequence of the orbits $\varphi_{t_i}(x)$ possesses limit points. We shall prove asymptotic stability by contradiction and so we assume the absence of the point $x = 0$ from the limit points of $\varphi_{t_i}(x)$. In other words, we assume that as $t_n \longrightarrow \infty$, $\varphi_{t_n}(x) \longrightarrow l \neq 0$. Next we draw up a sequence of inequalities

$$W(\varphi_{t_n}(x)) > W(\varphi_{t_{n+1}}(x)) > ... > W(l) \tag{5.99}$$

from the nondecreasing condition of $W$ and note that continuity implies $W(\varphi_{t_n}(x)) \longrightarrow W(l)$. Since $l$ is not a point of equilibrium, we can say that the orbit of the limit point $\varphi_m(l)$ must be such that it is $W(\varphi_m(l)) < W(l)$ for any positive $m$. By continuity it must hold that $W(\varphi_{t_n+m}(x)) \longrightarrow W(\varphi_{t_n}(l)) < W(l)$. We can choose $m$ in such a way that for $n'$ obeying $t_{n'} > t_{n+m}$

$$W(\varphi_{t_{n'}}(x)) < W(\varphi_{t_{n+m}}(x)) < W(l) \tag{5.100}$$

which runs counter to the previous inequality. So all the limit points of the flow are at the origin and $x = 0$ is asymtotically stable.

With $l$ the limit point, for large $n$, $x(t_n)$ can be made arbitrarily close to $l$.

We now turn to some examples to determine the Lyapunov function and then examine the stability according to the criterion prescribed above.

**Example 5.10**

The governing equation of the linear harmonic oscillator problem can be expressed as

$$\dot{x} = y, \quad \dot{y} = -\omega_0^2 x \tag{5.101}$$

It turns out that the energy function $E(x,y) = \frac{1}{2}\omega_0^2 x^2 + \frac{1}{2}y^2$ doubles up as the Lyapunov function. Indeed identifying it with $W(x,y)$, we notice that $W(0,0) = 0$ at the point of the minimum $(0,0)$ which is the equilibrium point. Further, in the region in the deleted neighborhood of $(0,0)$, $W(x,y) > 0$. Working out the time derivative of $W(x,y)$ we easily find

$$\dot{W} = \dot{x}\frac{\partial W}{\partial x} + \dot{y}\frac{\partial W}{\partial y} = y\omega_0^2 x - \omega_0^2 xy = 0 \tag{5.102}$$

So we conclude that the system is Lyapunov stable.

**Example 5.11**

Let us take up the damped oscillator problem already considered earlier in terms of two first order equations

$$\dot{x} = y, \quad \dot{y} = -\omega_0^2 x - \lambda\omega_0 y, \quad \lambda > 0 \tag{5.103}$$

We choose the Lyapunov function to be $W(x,y) = \frac{1}{2}\omega_0^2 x^2 + \frac{1}{2}y^2$. Since $W(x,y)$ is a sum of two squares, a unique minimum occurs at $(0,0)$ which is the equilibrium point. For the quantity $\nabla W(\vec{x}) \cdot \vec{\xi}$, we have $\frac{\partial W}{\partial x} = \omega_0^2 x$ and $\frac{\partial W}{\partial y} = y$ and $\vec{\xi}$ has components $y$ and $-\omega_0^2 x - \lambda\omega_0 y$. Substituting them we find $\nabla W(\vec{x}) \cdot \vec{\xi} = -\lambda\omega_0 y^2$ which is a strictly negative quantity for all $y$ since $\lambda > 0$. So the system is not only stable but asymptotically stable too.

**Example 5.12**

Let us address the pendulum problem as described by the equation of motion $\ddot{\theta} + \frac{g}{l}\sin\theta = 0$. It can also be represented in the following form:

$$\dot{\theta} = y, \quad \dot{y} = -\frac{g}{l}\sin\theta \tag{5.104}$$

We already know its fixed points to be $(n\pi.0)$, $n = 0, 1, 2, ....$ Consider the following candidate for the Lyapunov function:

$$W(\theta, y) = (1 - \cos\theta)gl + \frac{l^2 y^2}{2} \tag{5.105}$$

Its features are (i) $W(0,0) = 0$, (ii)$W(\theta, y) > 0$ for $-2\pi < \theta < 2\pi$ for $(\theta, y) \neq (0,0)$ and (iii) $\dot{W} = gl\dot{\theta}\sin\theta + l^2 y\dot{y} = 0$ when the above expressions for $\dot{\theta}$ and $\dot{y}$ are substituted. Hence we notice that the system is Lyapunov stable but not asymptotically so because $\dot{W}$ is not strictly negative for all $(\theta, y) \neq (0,0)$.

## 5.7 Van der Pol oscillator and limit cycles

Van der Pol oscillator is a mechanical model of a nonlinear system through the inclusion of a friction-like term with varying coefficient. We already encountered in Chapter 1 the case of a damped oscillator in which the friction term was accounted for both linearly and nonlinearly. In the Van der Pol case the governing equation reads

$$\ddot{x} + af(x, \dot{x}) + \omega_0^2 x = 0 \tag{5.106}$$

where the function $f(x.\dot{x}) = (x^2 - 1)\dot{x}$ incorporates the effects of friction. If the coupling $a = 0$ one is left with the harmonic oscillator equation. It is a second-order autonomous equation. The presence of $a$ renders the system non-conservative. It is to be noted that in the range $0 < x < 1$ the oscillator receives an additional impetus for the acceleration while for $1 > x > 0$ the volatility of the oscillator is reduced. Physically there is a contrasting role of the energy: it is generated in low amplitudes while large ones dissipate it. An interesting situation arises at some intermediate point when some kind of a compromise is reached between energy generation and dissipation resulting in oscillations forming around that state. The convergence toward such a state is characterized as a limit cycle. Limit cycles appear in phase spaces of two or more dimensions. The concept of an attractor was introduced earlier in the context of trajectories in a dynamical system approaching the final state in the phase space while evolving in time. A limit cycle is a second type of attractor exhibiting steady closed oscillations: it is an isolated closed curve which is stable or unstable depending upon whether the neighboring trajectories are attracted upon it or get diverged away. There is also another class of limit cycles called semi-stable limit cycles which attract nearby trajectories from one side but repel those on the other. Stable and unstable limit cycles are illustrated in Figure 5.9 and Figure 5.10.

To factorize (5.106) into a pair of first order equations, the usual trick is to set

$$x = r\cos(t + \phi), \quad \dot{x} = -r\sin(t + \phi) \tag{5.107}$$

where it is to be noted that we have introduced two time-dependent quantities $r$ and $\phi$ which are, respectively, the amplitude and the phase. The second equation does not represent the time derivative of the first but in order to be so we evaluate the time derivative of the first and look for consistency with the second. This yields the condition

$$\dot{r}\cos(t + \phi) - r\dot{\phi}\sin(t + \phi) = 0 \tag{5.108}$$

On the other hand if we take the time derivative of the second equation in (5.107) and compare with (5.106) we arrive at the requirement

$$\dot{r}\sin(t + \phi) + r\dot{\phi}\cos(t + \phi) + af = 0 \tag{5.109}$$

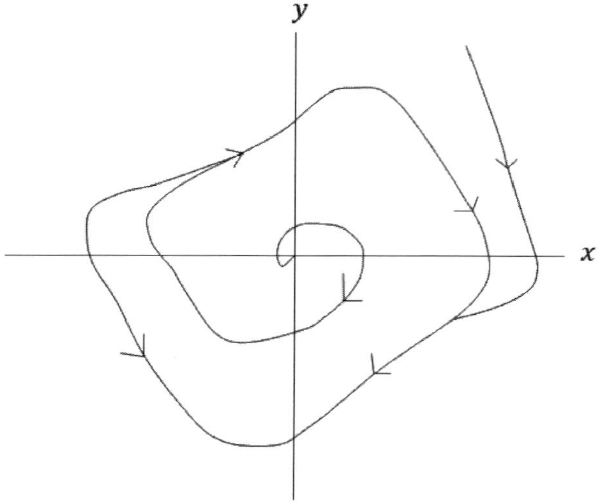


**FIGURE 5.9**: Stable limit cycle.

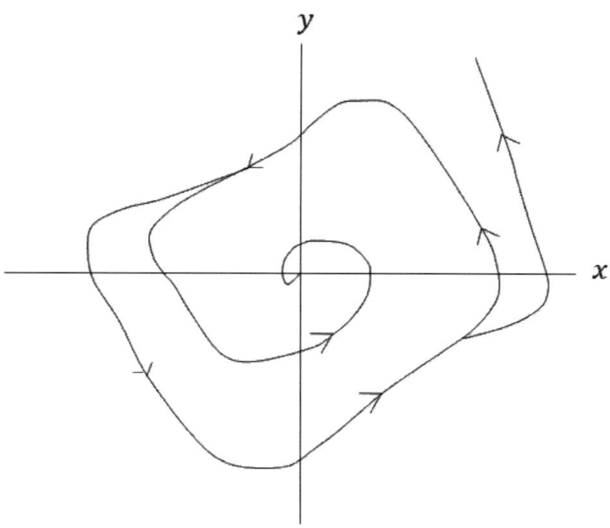


**FIGURE 5.10**: Unstable limit cycle.

where we have set $\omega_0 = 1$ for simplicity. It is now easy to solve for $\dot{r}$ and $\dot{\phi}$ from (5.108) and (5.109) by first multiplying the two equations by $\cos(t+\phi)$ and $\sin(t+\phi)$ and subtracting and then multiplying the two equations by $\sin(t+\phi)$ and $\cos(t+\phi)$ and adding to obtain, respectively,

$$\dot{r} = -a\sin(t+\phi)f, \quad \dot{\phi} = -\frac{a}{r}\cos(t+\phi)f \tag{5.110}$$

Equations (5.110) are an exact mapped version of (5.106) in terms of the variables $r$ and $\phi$. So it has indeed been possible to factorize (5.20). In the von der Pol case a specific form of $f$ is used which is

$$f(r,t) = -(1 - r^2\cos^2 t)r\sin t \tag{5.111}$$

We now explore the situation when the quantities $r$ and $\phi$ are slowly varying functions of time as is indeed the case when the system approaches the limit cycle. The averaged amplitude and phase are then given by

$$\frac{d\bar{r}}{dt} = -\frac{a}{2\pi}\int_0^{2\pi} f(r,\tau)\sin\tau d\tau, \quad \frac{d\bar{\phi}}{dt} = -\frac{a}{\bar{r}}\int_0^{2\pi} f(r,\tau)\cos\tau d\tau \tag{5.112}$$

where the integration is taken over the period of oscillation.

Using the form (5.111) for $f$ the integrations are straightforward to perform using the well-known results

$$\int_0^{2\pi}\cos^2\tau d\tau = \frac{1}{2}, \quad \int_0^{2\pi}\sin^2\tau\cos^2\tau d\tau = \frac{1}{8}, \quad \text{and} \quad \int_0^{2\pi}\sin^2\tau\sin\tau\cos\tau d\tau = 0 \tag{5.113}$$

We obtain

$$\frac{d\bar{r}}{dt} = a\frac{\bar{r}}{2}(1 - \frac{\bar{r}^2}{4}), \quad \frac{d\bar{\phi}}{dt} = 0 \tag{5.114}$$

The $\bar{r}$ integration is easily performed by setting $\bar{r} = u^{-\frac{1}{2}}$ yielding the form

$$\bar{r}(t) = \frac{2e^{\frac{at}{2}}}{\sqrt{4 + r_0^2 e^{at} - r_0^2}} \tag{5.115}$$

where $r_0$ is given an initial condition.

It is worthwhile to inquire into the nature of the fixed points of the first equation in (5.114). By inspection the fixed points are at $\bar{r} = 0$ and $\bar{r} = 2$. Performing a slight perturbation around $\bar{r} = 0$, i.e., setting $\bar{r} = 0 + \delta$, where $\delta$ is a small quantity gives

$$\dot{\delta} = \frac{a}{2}\delta + O(\delta^3) \tag{5.116}$$

Therefore $\delta$ goes as $e^{\frac{at}{2}}$ which blows up asymptotically. If we do the same around $\bar{r} = 2$, i.e., set $\bar{r} = 2 + \delta$ the situation turns out to be entirely different. We find

$$\dot{\delta} = -\frac{a}{2}\delta + O(\delta^2) \tag{5.117}$$

Therefore $\delta$ goes as $e^{-\frac{at}{2}}$ signalling $\bar{r} = 2$ to be a stable equilibrium point.

Now concerning the solution (5.115) it is evident for large $t$ it approaches the circle of constant radius 2 from both the inner side and the outer side of the radius for all values of the initial condition $r_0$. We therefore identify it as a limit cycle. The critical value of $\bar{r} = 2$ which is a point of stable equilibrium has thus an interesting feature of attracting other points in the phase space as time evolves. Limit cycles are not too infrequent in nature. Consider the following example.

**Example 5.13**

Show that the following nonlinear system given by the pair of equations

$$\dot{x} = y + \frac{x}{\sqrt{x^2 + y^2}}[1 - (x^2 + y^2)], \quad \dot{y} = -x + \frac{y}{\sqrt{x^2 + y^2}}[1 - (x^2 + y^2)]$$

has a limit cycle.

Transforming to polar coordinates by setting $x = r\cos\theta$ and $y = r\sin\theta$ gives the following forms in the $(r, \theta)$ plane:

$$\dot{r} = 1 - r^2, \quad \dot{\theta} = -1$$

On integrating we find

$$r(t) = \frac{ce^{2t} - 1}{ce^{2t} + 1}, \quad \theta(t) = \theta_0 - t$$

where $c = \frac{1+r_0}{1-r_0}$ and $(r_0, \theta_0)$ are a set of initial conditions. We readily notice that as $t$ increases $r$ approaches a circle of radius 1 with constant $\theta$-value. In particular, when $r > 1$ it is seen that $\dot{r} < 0$ while for $r < 1$ it is found that $\dot{r} > 0$. In the former case the trajectories spiral clockwise inward while in the latter case the trajectories spiral clockwise outward. Thus, $r(t)$ is a clear case of a limit cycle.

Limit cycles are an intrinsic property of nonlinear systems. However, limit cycles can also arise by piecing together linear equations. We follow the treatment presented in [2]. We consider the damped oscillator problem on two sides of the zero-boundary of the velocity variable given by the following pair of linear equations:

$$\ddot{x} + 2g\dot{x} + \omega_0^2 x = 0 \quad \dot{x} < 0 \tag{5.118}$$

$$\ddot{x} + 2g\dot{x} + \omega_0^2 (x - \lambda) = 0 \quad \dot{x} > 0 \tag{5.119}$$

where $\lambda$ is a real constant. We also restrict $g$ to the interval $0 < g < \omega_0$.

While the general solution of (5.118) is given by

$$x(t) = Ae^{-gt}\cos(\beta t + \phi) \tag{5.120}$$

where $\omega_0^2 = \beta^2 + g^2$, the one for (5.119) has the form

$$x(t) = \lambda + Be^{-gt}\cos(\beta t + \psi) \tag{5.121}$$

where $A$ and $B$ are the amplitudes.

Projecting out (5.119) in an equivalent form

$$\dot{x} = y, \quad \dot{y} = -\bar{\omega_0}^2(x - \lambda) - 2gy \tag{5.122}$$

it is easily seen that the point $(\lambda, 0)$ is a stable focus.

Let us now follow the motion in the phase plane defined by the coordinates $(x, y)$. Consider the lower half plane. Here a typical point will be guided by the equation (5.118). If the motion starts from say, the point $(x_1, 0)$ on the $x$-axis at time say, $t_1$, then it obeys according to

$$x_1(t_1) = Ae^{-gt_1}\cos(\beta t_1 + \phi) \equiv x_1 \tag{5.123}$$

When the trajectory again cuts the $x$-axis moving in the clockwise direction at the point after a lapse of time say, $(t_1 + \frac{\pi}{\beta})$, (5.120) this time gives

$$x_1(t_1 + \frac{\pi}{\beta}) = Ae^{-g(t_1+\frac{\pi}{\beta})}\cos(\beta t_1 + \pi + \phi) = -Ae^{-g(t_1+\frac{\pi}{\beta})}\cos(\beta t_1 + \phi) \tag{5.124}$$

Comparison of (5.123) and (5.124) shows that the factor $Ae^{-gt_1}\cos(\beta t_1 + \phi)$ can be canceled out leaving us with the relation

$$x_1(t_1 + \frac{\pi}{\beta}) = -\varsigma x_1 \tag{5.125}$$

where the coefficient factor $\varsigma (\equiv e^{-\frac{g\pi}{\beta}})$ is smaller than unity because $\frac{g}{\beta} = \frac{g}{\sqrt{(\omega_0 - g)(\omega_0 + g)}}$ and $0 < g < \omega_0$.

Next, as the representative point $x_1$ enters the upper half plane, it comes under the influence of Equation (5.119). The relevant equations as the point travels from $(t_1 + \frac{\pi}{\beta})$ until it meets the $x$-axis again at $(t_1 + \frac{2\pi}{\beta})$ are

$$x_1(t_1 + \frac{\pi}{\beta}) - \lambda = Be^{-g(t_1+\frac{\pi}{\beta})}\cos(\beta t_1 + \pi + \psi) = -Be^{-g(t_1+\frac{\pi}{\beta})}\cos(\beta t_1 + \psi) \tag{5.126}$$

$$x_1(t_1 + \frac{2\pi}{\beta}) - \lambda = -B\varsigma e^{-g(t_1+\frac{\pi}{\beta})}\cos(\beta t_1 + \pi + \psi) = B\varsigma e^{-g(t_1+\frac{\pi}{\beta})}\cos(\beta t_1 + \psi) \tag{5.127}$$

Utilizing (5.125) a combination of the above two equations gives

$$x_1(t_1 + \frac{2\pi}{\beta}) = \lambda(1 + \varsigma) + \varsigma^2 x_1 \tag{5.128}$$

Relation (5.128) can now be iterated to write down the coordinates of the chain of the intersections of the particle trajectory with the positive $x$-axis in terms of the starting coordinate $x_1$ only as the particle alternately moves in the lower and upper half phase plane:

$$x_1, \lambda(1+\varsigma^2)+\varsigma^2 x_1, \lambda(1+\varsigma^2+\varsigma^3)+\varsigma^4 x_1, ..., \lambda(1+\varsigma^2+...+\varsigma^{2N-1})+\varsigma^{2N} x_1, ... \tag{5.129}$$

where $N$ denotes the number of complete revolutions. Since $\varsigma < 1$, $\varsigma^{2N} x_1 \to 0$ as the number of revolutions grow. So the coefficient of $\lambda$ is left with the form of a geometric series. The series of intersecting points thus converges to the limit $\frac{\lambda}{1-\varsigma}$.

To understand the significance of this result we observe that in Equation (5.128) the criterion of the closed orbit would imply $x_1(t_1 + \frac{2\pi}{\beta}) = x_1(t_1)$ resulting in the unique solution $x_1 = \frac{\lambda}{1-\varsigma}$. It is exactly the convergence point of the geometric series as just now noticed. Thus our finding is that all the trajectories asymptotically approach a unique attracting limit cycle.

Poincare–Bendixson theorem which is often found to be useful in the context of a limit cycle is of much use in dynamical systems. It is essentially a two-dimensional result. It states (and we state it without a proof) that if a flow is confined within a closed and bounded domain $D$ which does not contain any fixed point, then $D$ holds at least one periodic orbit which is a limit cycle. Note that it speaks only of sufficient conditions. For practical utility of this theorem, one needs to find a trapping region $D$ containing no fixed point and in which trajectories enter but do not move out. The following example will make this issue and also the applicability of Poincare–Bendixson theorem clear.

**Example 5.14**

Show that a physical system given by the equation $\ddot{x}-(1-3x^2-2\dot{x}^2)+x=0$ has a limit cycle.

We can express the above system as

$$\dot{x} = y, \quad \dot{y} = y(1-3x^2-2\dot{x}^2) - x$$

It transpires by putting equal to zero the right-hand sides that the origin $(0,0)$ is the only fixed point. Transforming to polar coordinates by setting $x = r\cos\theta$ and $y = r\sin\theta$ gives the following forms of the equations in the $(r,\theta)$ plane

$$\dot{r} = r(1-2r^2-r^2\cos^2\theta)\sin^2\theta, \quad \dot{\theta} = -1+(1-3r^2\cos^2\theta-2r^2\sin^2\theta)\sin\theta\cos\theta$$

This implies that $\dot{r} > 0$ for $1 > 2r^2 + r^2\cos^2\theta > 0$ or for

$$r^2 < \frac{1}{2+\cos^2\theta} < \frac{1}{3}$$

while $\dot{r} < 0$ for $1 < 2r^2 + r^2\cos^2\theta > 0$ or for

$$r^2 > \frac{1}{2+\cos^2\theta} > \frac{1}{2}$$

Comparing the two bounds we arrive at the definition of a domain $D$ given by

$$D = \{(r,\theta) : \frac{1}{\sqrt{3}} < r < \frac{1}{\sqrt{2}}\}$$

in which there is no fixed point, the latter existing only at the origin as observed earlier. $D$ thus corresponds to an annulus or the trapping zone which excludes the fixed point $(0,0)$ and is bounded by the circles of radii $\frac{1}{\sqrt{3}}$ and $\frac{1}{\sqrt{2}}$. Furthermore, this domain corresponds to a region for which $\dot{r} > 0$ on the one side but $\dot{r} < 0$ on the other. This means that a trajectory once inside $D$ stays within it and cannot escape. By Poincare–Bendixson theorem there is a limit cycle in $D$.

---

## 5.8 Bifurcations

The change in the nature of the fixed point leading to a drastic change in the behavior of the trajectories in the neighborhood of the fixed point gives rise to bifurcation. Let us have a look at the damped oscillator problem considered earlier in this chapter. We found a change in the nature of the fixed point from a stable focus to a stable node as $\lambda$ moved out from the interval $0 < \lambda < 2$ to the region $\lambda > 2$ . Since stability persists we cannot call $\lambda = 2$ to be a bifurcation point. However, the other fixed point namely $\lambda = 0$, is indeed a bifurcation point since there is a sudden change in the nature of stability: while at $\lambda = 0$, the real part of the eigenvalue vanishes and we are left with a center at the origin, for $\lambda < 0$, there is an unstable focus and unstable node for the fixed point $(0,0)$ corresponding to the respective cases of $\lambda^2 < 4$ and $\lambda^2 \geq 4$.

Bifurcation is thus concerned with the qualitative sudden changes with the model, namely, the phase portraits changing in a non-continuous way, as the underlying parameters controlling the system undergo variation. Bifurcations play a leading role in the theory of nonlinear dynamics. A complete treatment of such a subject is beyond the scope of this book. We give a few illustrative examples to show how bifurcations occur in physical systems. To this end we consider four types of bifurcations:

(i) Saddle-node bifurcation
(ii) Transcritical bifurcation
(iii) Pitchfork bifurcation
(iv) Hopf bifurcation

The first three types deal with real eigenvalues while in the fourth one complex eigenvalues are relevant.

We consider an autonomous differential equation

$$\frac{dx}{dt} = f(x, \mu), \quad x \in R, \quad \mu \in R \tag{5.130}$$

involving some parameter $\mu$. To detect bifurcation points we write down the linearized flow and search for those parameter values for which the fixed point has a zero or purely imaginary eigenvalue:

$$\text{either} \quad \text{Re} \quad \beta = 0 \quad \text{or} \quad \det M = 0 \tag{5.131}$$

where $M$ stands for the Jacobi matrix and $\beta$ denotes an eigenvalue. Note that it is only for non-hyperbolic fixed points that bifurcations can occur. In other words, we set

$$f(\bar{x}, \bar{\mu}) = 0, \quad \frac{\partial f(x, \mu)}{\partial x}|_{(\bar{x}, \bar{\mu})} = \beta = 0 \tag{5.132}$$

where $\bar{x}$ is a bifurcation point and $\bar{\mu}$ is the critical value of the parameter $\mu$ where the bifurcation arises.

We now turn to a discussion of the four types of bifurcations as mentioned above.

### (i) Saddle-node bifurcation

A saddle-node bifurcation is characterized by the following features at the point $(\bar{\mu}, \bar{x})$:

$$\frac{\partial f}{\partial x} = \beta = 0, \quad \frac{\partial f}{\partial \mu} \neq 0, \quad \frac{\partial^2 f}{\partial x^2} \neq 0, \tag{5.133}$$

A typical model equation admitting a saddle-point bifurcation is

$$\frac{dx}{dt} = f(x) = \mu - x^2, \quad x \in R, \quad \mu \in R \tag{5.134}$$

For $\mu > 0$, the equilibrium or fixed points are at

$$\bar{x}_+ = \sqrt{\mu}, \quad \bar{x}_- = -\sqrt{\mu} \tag{5.135}$$

while for $\mu < 0$, there is no fixed point because of the reality of $x$. Of the two fixed points, it can be very easily checked that the one at $\sqrt{\mu}$ is a node and the one at $-\sqrt{\mu}$ is a repeller. What happens at the point $\mu = 0$? We notice that there is an abrupt jump in the number of fixed points as $\mu$ passes through the zero value and a bifurcation occurs at $\mu = 0$: the solution $\bar{x}_+ = \sqrt{\mu}$ is a stable one, in contrast, the other one corresponding to $\bar{x}_+ = -\sqrt{\mu}$ is unstable. The bifurcation diagram is shown in Figure 5.11.

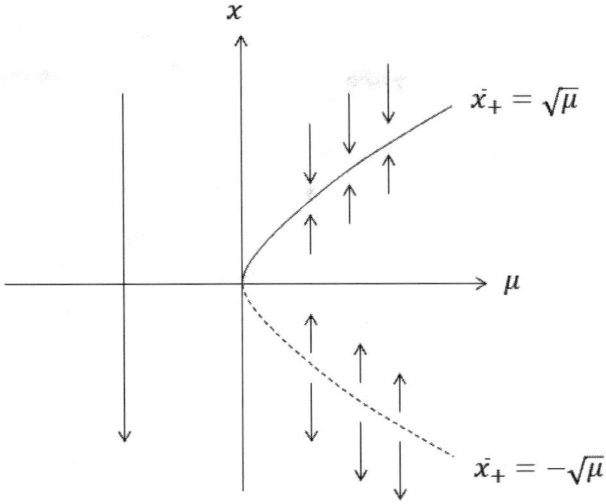


**FIGURE 5.11**: Saddle-node bifurcation. The solid line represents the stable portion and the dashed line represents the unstable portion.

### (ii) Transcritical bifurcation

A transcritical bifurcation has the following features at the point $(\bar{\mu}, \bar{x})$:

$$\frac{\partial f}{\partial x} = \beta = 0, \quad \frac{\partial^2 f}{\partial x^2} \neq 0, \quad \frac{\partial^2 f}{\partial x \partial \alpha} \neq 0, \tag{5.136}$$

Here a typical model equation admitting a transcritical bifurcation is

$$\frac{dx}{dt} = f(x) = \mu x - x^2, \quad x \in R, \quad \mu \in R \tag{5.137}$$

The equilibrium or fixed points are at the points 0 and $\mu$. To examine their characters we find the following features:

$$\mu < 0 : \bar{x} = \alpha(\text{unstable}), \quad \bar{x} = 0(\text{stable}) \tag{5.138}$$

$$\mu > 0 : \bar{x} = \alpha(\text{stable}), \quad \bar{x} = 0(\text{unstable}) \tag{5.139}$$

The two signs of $\alpha$ convey two opposite behaviors: both the fixed points 0 and $\alpha$ switch from being unstable(stable) to stable(unstable). Figure 5.12 gives the bifurcation diagram which is of transcritical type.

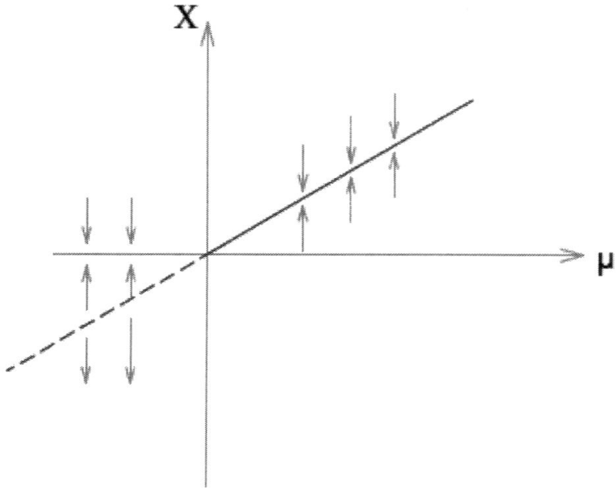


**FIGURE 5.12**: Transcritical bifurcation. The stable and unstable portions are distinguished by solid and dashed lines.

### (iii) Pitchfork bifurcation

The following features govern a pitchfork bifurcation. We have at the point $(\bar{\mu}, \bar{x})$:

$$\frac{\partial f}{\partial x} = \beta = 0, \qquad \frac{\partial^3 f}{\partial x^3} \neq 0, \qquad \frac{\partial^2 f}{\partial x \partial \alpha} \neq 0, \tag{5.140}$$

For a pitchfork bifurcation a typical model equation is

$$\frac{dx}{dt} = f(x) = \mu x - x^3, \quad x \in R, \quad \mu \in R \tag{5.141}$$

The equilibrium or fixed points are at

$$\bar{x}_+ = \sqrt{\mu}, \quad \bar{x}_- = -\sqrt{\mu}, \quad \bar{x}_0 = 0 \tag{5.142}$$

We immediately notice that for $\mu < 0$, the system has only one fixed point at $\bar{x}_0 = 0$. It is stable. On the other hand, for $\mu > 0$ the number of fixed points are two, namely, $\sqrt{\mu}$ and $-\sqrt{\mu}$, both of which are stable but in this case $\bar{x}_0 = 0$ is unstable. Figure 5.13 gives the supercritical bifurcation diagram which is of pitchfork type. It is to be observed that the curve for $\mu = 0$ simply passes through the origin and does not cut the abscissa anywhere else. However for a positive $\mu$ the cubic meets the abscissa at three points. The subcritical pitchfork bifurcation diagram, for which the model equation stands as $\frac{dx}{dt} = f(x) = \mu x + x^3$, is illustrated in Figure 5.14.

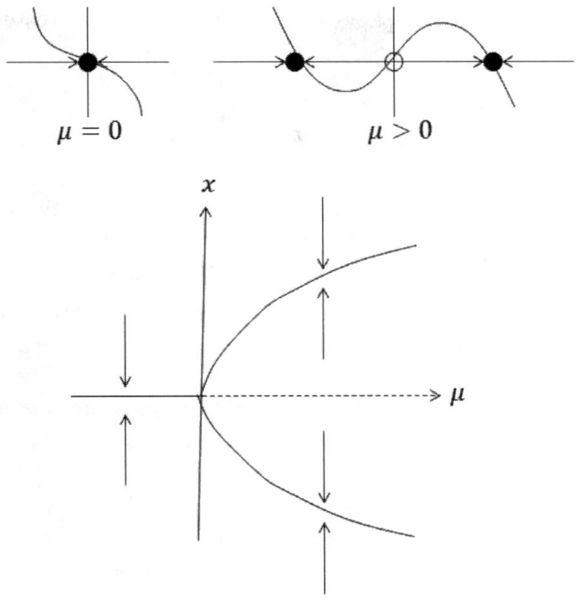


**FIGURE 5.13**: Supercritical pitchfork bifurcation.

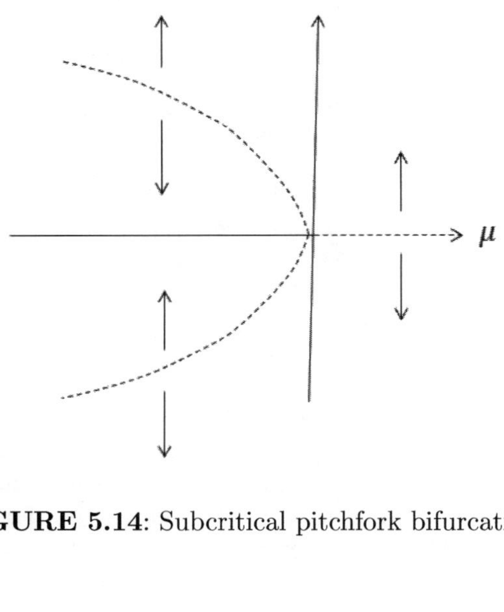


**FIGURE 5.14**: Subcritical pitchfork bifurcation.

### (iv) Hopf bifurcation

Hopf bifurcation constitutes an important class of bifurcations in nonlinear dynamics. It arises in situations when a complex pair of eigenvalues at a fixed point sheds its real part and becomes purely imaginary. Thus its occurrence has relevance for systems defined in phase space with dimension $m \geq 2$. In assessing its occurrence we are particulary interested in a theorem called Hopf bifurcation theorem whose basic formulation is concerned with sufficient conditions related to observation of bifurcations to cycles and the orbital stability of the latter. Without getting into the details of the underpinnings of the theorem, let us only note that if a differential equation of the autonomous type

$$\frac{d\vec{X}}{dt} = \vec{f}(\vec{X}, \mu) \tag{5.143}$$

is endowed with the conditions that the associated Jacobi matrix has an isolated fixed point at which a pair of conjugate-complex eigenvalues $\beta_{1,2}$ appear (whose appearence of course depends on some control parameter $\mu$ assuming a specific value) such that (i) at a critical value of $\mu$ say, $\bar{\mu}$, the real part vanishes but the imaginary part remains nonzero and (ii) at the same point $\bar{\mu}$ the derivative of the real part $\beta$ is nonzero as well, then the system of differential equations (5.143) admits periodic solutions which are both stable and unstable. Before we turn to the applicability of the theorem when the criterium of its conditions are fulfilled, we note that the nature of Hopf bifurcation can be supercritical or subcritical. In the supercritical case a stable limit cycle is formed around an unstable equilibrium point whereas in the subcritical case it is an unstable limit cycle that is formed around a stable equilibrium point.

Let us focus on the following examples. First take the Van der Pol oscillator equation

$$\ddot{x} + (x^2 - \mu)\dot{x} + x = 0 \tag{5.144}$$

where $\mu$ is a control parameter. Expressing it in the pair

$$\dot{x} = y, \quad \dot{y} = -(x^2 - \mu)y - x \tag{5.145}$$

the equilibrium point can be seen to be located at $(0, 0)$, the origin. The underlying Jacobi matrix for the linearized system is given by

$$M = \begin{pmatrix} 0 & 1 \\ -1 & \mu \end{pmatrix} \tag{5.146}$$

Its eigenvalues $\beta$ are easily determined which have a common real part $\frac{\mu}{2}$ and imaginary parts $\pm\frac{i}{2}\sqrt{4-\mu^2}$. The real part is seen to vanish at $\mu = 0$ at which point the imaginary part is nonzero: $\text{Im}\beta = -1$. Further the derivative of the real part with respect to $\mu$ at $\mu = 0$ is $\frac{1}{2}$ which is nonzero. Hence the two conditions of Hopf bifurcation theorem are satisfied. Now, as just noted, the real part of the eigenvalues is positive for $\mu > 0$ and negative for $\mu < 0$. So the equilibrium is an unstable focus for $\mu > 0$ but stable for $\mu < 0$. However, there is a stable periodic orbit (limit cycle) if $\mu > 0$ for the system (5.144) (see Exercise 10) and hence our conclusion is that at $\mu = 0$ there is a supercritical bifurcation.

Another case of supercritical bifurcation occurs for the system

$$\dot{r} = r(\mu - r^2), \quad \dot{\theta} = 1, \quad r \geq 0 \tag{5.147}$$

The Cartesian version of the above equation reads

$$\dot{x} = \mu x - y - x(x^2 + y^2), \quad \dot{y} = x + \mu y - y(x^2 + y^2) \tag{5.148}$$

which straightforwardly yields (5.49) by transferring to polar coordinates $x = r\cos\theta$ and $y = r\sin\theta$.

The only equilibrium point for the above system is at the origin $(0, 0)$, i.e., at $r = 0$. Its behaviors are (i) for $\mu > 0$, an unstable focus (complex eigenvalues having a positive real part), (ii) for $\mu < 0$, a stable focus (complex eigenvalues having a negative real part) and (iii) for $\mu = 0$, a center (eigenvalues are purely imaginary). For $\mu < 0$, since the governing equation shows $\dot{r} < 0$, it indicates that the trajectories spiral asymptotically toward the origin in an anticlockwise manner due to the positive sign of $\dot{\theta}$. However, for $\mu > 0$, $\dot{r}$ changes sign as one steps out of the interval from $(0, \sqrt{\mu})$ to $(\sqrt{\mu}, \infty)$, there being a stable limit cycle at $r = \mu$. Since the origin is an unstable focus we encounter a supercritical Hopf bifurcation.

We now provide an example of a subcritical bifurcation. Consider a system which is given by

$$\dot{r} = \mu r + 2r^3 - r^5, \quad \dot{\theta} = 1, \quad r \geq 0 \tag{5.149}$$

In Cartesian coordinates the equivalent form is

$$\dot{x} = \mu x - y - x(x^2 + y^2)(2 - x^2 - y^2), \quad \dot{y} = x + \mu y + y(x^2 + y^2)(2 - x^2 - y^2) \tag{5.150}$$

For small values of $r$ it is clear that the equilibrium point exists at the origin $(0, 0)$, i.e., at $r = 0$: it is a stable focus for $\mu < 0$ but an unstable one for $\mu > 0$. For $\mu = 0$, however, the origin is unstable.

Let us focus on the interval $-1 < \mu < 0$. The radial equation admits of a factorization in the following form:

$$\dot{r} = -r(r - \sqrt{1 + z})(r + \sqrt{1 + z})(r - \sqrt{1 - z})(r + \sqrt{1 - z}) \tag{5.151}$$

where $z = \sqrt{1 + \mu}$. Note that in the interval $-1 < \mu < 0$, both $1 + z$ and $1 - z$ are positive quantities. We therefore identify two limit cycles of radii $r_1 = \sqrt{1 + z}$ and $r_2 = \sqrt{1 - z}$. Further, $\dot{r}$ changes sign as follows: it is negative in $0 < r < r_1$, positive in $r_1 < r < r_2$ and again negative for $r > r_2$. We therefore observe that the limit cycle $r_1 = \sqrt{1 - z}$ is unstable but $r_2 = \sqrt{1 + z}$ is stable.

Turning to the case $\mu > 0$ now, the factorization is to be carried out somewhat differently, namely,

$$\dot{r} = -r(r - \sqrt{1+z})(r + \sqrt{1+z})(r^2 + z - 1) \tag{5.152}$$

The reason is that in this case while $1 + z$ is positive, the quantity $1 - z$ is negative. This means that we only have one limit cycle of radius $r_2 = \sqrt{1+z}$. Concerning the signs of $\dot{r}$, we see that it is positive in the interval $0 < r < r_2$ but negative for $r > r_1$. Hence it is stable. We therefore have a subcritical Hopf bifurcation because in the interval $-1 < \mu < 0$ the origin which is a stable focus has around it an unstable limit cycle of radius $r_1 = \sqrt{1-z}$ and a stable limit cycle of radius $r_2 = \sqrt{1+z}$ whereas for $\mu > 0$ it is the origin which is an unstable focus but has around it a stable limit cycle of radius $r_2 = \sqrt{1+z}$.

---

## 5.9 Summary

In this chapter we conducted a broad survey of some of the major issues of a dynamical system. As is widely recognized, a modern treatment of classical mechanics is incomplete without an insight from the standard ideas and techniques of dynamical systems. Among them are essentially the theory of stability, limit cycles and bifurcations. (We refrained from giving any treatment on chaos which is beyond the scope of this book). To approach these ideas we first provided the definitions of various concepts that frequently appear in dynamical systems. We placed a greater emphasis on the linear systems and discussed how the process of linearization can be carried out for different classes of nonlinear systems under certain suitable conditions. In this regard many guided examples were worked out. We discussed the Lotka–Volterra model which is a popular model for prey-predator species. We considered stability of solutions and the role of the Lyapunov function. The concept of limit cycles was explained by means of models including the one of Van der Pol oscillator. Several aspects of bifurcations were treated through model examples.

# Exercises

1. Show that the origin of the system

$$\dot{x} = -x - y^3 - x^2 - y^2, \quad \dot{y} = x - y + x^2 + y^2$$

is a stable focus.

2. Obtain the Lyapunov function for the system

$$\dot{x} = xy - y^3 - x^3 - y^2, \quad \dot{y} = xy - x^2$$

What is the character of the origin?

3. Noting that a linear harmonic oscillator equation can be written as a pair $\dot{x} = y, \quad \dot{y} = -\omega_0^2 x$, examine the stability and asymptotic stability of the modified system $\dot{x} = y, \quad \dot{y} = -\omega_0^2 x - \alpha y^3(1 + x^2)$ when $\alpha >$ or $< 0$.

4. Determine the fixed points and analyze the stability properties of the Duffing oscillator

$$\ddot{x} + \dot{x} - x + x^3 = 0$$

5. For the Brusselator model

$$\dot{x} = 1 - (1 + b)x + x^2 y, \quad \dot{y} = bx - x^2 y$$

show by carrying out an analysis based on the Jacobian determinant method for $b < 2$ there is a stable spiral mode where in the range $2 < b < 4$ we have a spiral repellor.

6. Find an appropriate Lyapunov function for the system

$$\dot{x} = -x + 2y^3 - 2y^4, \quad \dot{y} = -x - y + xy$$

to show that the origin is an asymptotically stable point.

7. Show for the system

$$\dot{x} = -y + x(1 - x^2 - y^2), \quad \dot{y} = x + y(1 - x^2 - y^2)$$

there exists an annulus bounded by the circles of radii $\frac{1}{2}$ and 2 that contains no fixed point. By the help of Poincare–Bendixson theorem deduce that there is at least one limit cycle in this annulus.

8. Determine the equation of the flow for the above equation in polar coordinates in the form

$$\varphi_t(r, \theta) = [(\frac{r^2 e^{2t}}{1 - r^2 + r^2 e^{2t}})^{\frac{1}{2}}, \theta + t]]$$

and also show that the unit circle describes a periodic orbit with period $2\pi$.

9. Use the Poincare–Benedixson theorem to show that the system

$$\dot{x} = y + \frac{1}{4}x(1 - 2x^2 - 2y^2), \quad \dot{y} = -x + \frac{1}{2}y(1 - x^2 - y^2)$$

admits a limit cycle inside the trapping region $\frac{1}{2} \leq r \leq 2$. Show that there is a smaller region given by $\frac{1}{\sqrt{2}} \leq r \leq 1$ containing the orbit.

10. Show that a stable periodic orbit exists for the system

$$\ddot{x} + (x^2 - \mu)\dot{x} + x = 0$$

11. Show that the Rayleigh equation

$$\ddot{x} + (x^2 - \mu)\dot{x} + x = 0$$

can be converted to the Van der Pol form by differentaiting it with respect to time and putting $y = \dot{x}$. Show that both the equations have the same first-order representation.

12. Analyze the following equations with regard to their bifurcation properties:

$$(i)\dot{x} = f(x) = 1 + \mu x + x^2$$

$$(ii)\dot{x} = f(x) = \mu - \cosh x$$

$$(iii)\dot{x} = f(x) = x - \mu x(1 - x)$$

$$(iv)\dot{x} = f(x) = \mu x - 4x^3$$

$$(v)\dot{x} = f(x) = x + \frac{\mu x}{1 + x^2}$$

13. Study the pitchfork bifurcation for the system

$$\dot{x} = \mu x + y + \sin x, \quad \dot{y} = x - y$$

where $\mu$ is a parameter.

14. Analyze the system

$$\dot{x} = \mu x - y - (x + \frac{3}{2}y)(x^2 + y^2), \quad \dot{y} = x + \mu y + (\frac{3}{2}x - y)(x^2 + y^2)$$

for Hopf bifurcation.

15. Consider the Brusselator model involving an additional parameter $\mu$

$$\dot{x} = \mu - (1 + b)x + x^2 y, \quad \dot{y} = bx - x^2 y$$

Discuss the stability of the fixed points and the existence of a limit cycle. Comment on the Hopf bifurcation if $\mu = 1$.

# Chapter 6

# Action principles



An action principle is basically a variational principle with the central idea derived by defining a functional, for two states of a physical system, corresponding to their initial and final configurations as the system evolves in time, which yields a stationary value when taken over the actual path which links these configurations, as compared to the neighboring varied paths having the same configurations, provided the total energy remains the same in the varied motion as in the actual motion. In simple terms, the action is defined in terms of the time integral of the Lagrangian between two fixed values of time that are identified with the initial and final position of the particle during the course of its motion. Action principles are of paramount importance in classical mechanics in that almost all formulations of physics admit an action principle interpretation while conversely any formalism resulting from an action principle is termed as well defined.We first introduce the principle of stationary action.

## 6.1 The principle of stationary action

Consider a physical system with $n$ degrees of freedom. Let $P$ denote a point particle described by a set of $n$ generalized coordinates $q_1, q_2, ..., q_n$. We call it the representative point of the system. As the system evolves, the coordinates $q_1, q_2, ..., q_n$ change with time and $P$ moves in the configuration space as dictated by the forces acting upon it. As shown in Figure 6.1, $C$ denotes the actual path between points $P_0$ and $P_1$ along which the motion

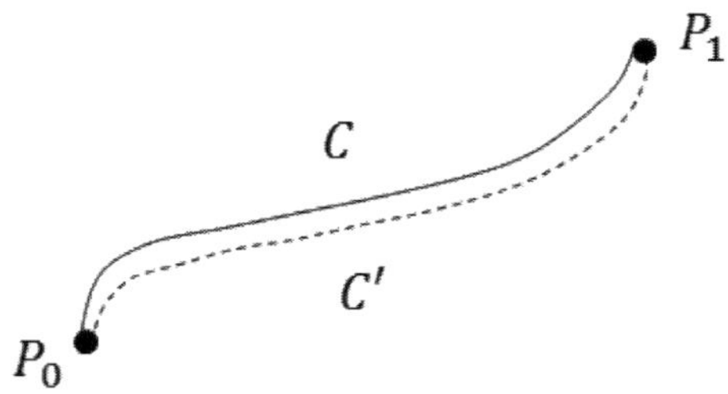


**FIGURE 6.1**: Two adjecent trajectories between representative points $P_0$ and $P_1$.

is executed. $P_0$ and $P_1$ correspond to the representative points of the system at times $t = t_0$ and $t_1$, respectively. As the system evolves the representative point traces out the curve $C$ joining the points $P_0$ and $P_1$. The curve is called the *trajectory* of the system. The actual path is the one all along which the equations of motion are satisfied. Thus $C$ is a dynamically allowed path and is distinguished from the neighboring ones (which may even lie infinitesimally close to $C$) in that these are only geometrically possible but dynamically impossible. In other words, the values of the coordinates and velocities that could exist on the curve $C'$ will not satisfy the equations of motion defined $C$.

We already encountered the notion of virtual variations $\delta q_i$ which are consistent with the conditions of the constraints. Note that $\delta$ is a shorthand notation for the variation. Let $C'$ be a geometrically allowed path which is virtually displaced from the actual path $C$. Recall that in a virtual displacement *no passage of real time* is involved. While the actual path is traced out by the coordinate $q_i$, the neighboring virtual path can be specified by the displaced coordinate $q_i + \delta q_i$. The corresponding velocities for the actual and virtual paths would be, respectively, $\dot{q}_i$ and $\dot{q}_i + \delta \dot{q}_i$. The comparison of the two paths, the actual and virtual, is, of course, done at the same point of time. Since

$$\frac{d}{dt}(q_i + \delta q_i) = \dot{q}_i + \frac{d}{dt}(\delta q_i)$$

we readily identify

$$\frac{d}{dt}(\delta q_i) = \delta \dot{q}_i \tag{6.1}$$

from the remark just made. It is then implied that the time derivative operator and the variation operation are interchangeable bringing out the commutative character of these two quantities: $d\delta = \delta d$.

We now look into the consequences of (6.1) for the change in the Lagrangian. We find

$$\begin{aligned}\delta L &= \sum_{i=1}^{n}\left[\frac{\partial L}{\partial q_i}\delta q_i + \frac{\partial L}{\partial \dot{q}_i}\delta \dot{q}_i\right] \\ &= \sum_{i=1}^{n}\left[\frac{\partial L}{\partial q_i} - \frac{d}{dt}\left(\frac{\partial L}{\partial \dot{q}_i}\right)\right]\delta q_i + \frac{d}{dt}\sum_{i=1}^{n}\frac{\partial L}{\partial \dot{q}_i}\delta q_i \\ &= \frac{d}{dt}\sum_{i=1}^{n}\frac{\partial L}{\partial \dot{q}_i}\delta q_i \end{aligned} \tag{6.2}$$

where we have used Lagrange's equations of motion which hold on the actual trajectory. The appearance of total time derivative in the right-hand side of (6.2) implies that if we integrate from an initial time $t_0$ to the final time $t_1$, then

$$\int_{t_0}^{t_1} \delta L dt = \sum_{i=1}^{n} \frac{\partial L}{\partial \dot{q}_i}\delta q_i \,|_{t_0}^{t_1} \tag{6.3}$$

Now writing

$$S[q_i] = \int_{t_0}^{t_1} L(q_i, \dot{q}_i, t)dt, \quad i = 1, 2, ..., n \tag{6.4}$$

we immediately notice that $\delta S$ corresponds to the left-hand side of (6.3). This is because due to (6.1) we can write

$$\delta S = \int (\delta L)dt + \int L\delta(dt) = \int (\delta L)dt + \int Ld(\delta t) = \int \delta L dt$$

since for a virtual variation $\delta t = 0$.

The action[1] of the path is given by the functional $S[q_i]$: it is the time integral of the Lagrangian between the terminal values of $t_0$ and $t_1$ along a particular path $q_i(t)$. We can represent (6.3) as

$$\delta S = \sum_{i=1}^{n} p_i \delta q_i \,|_{t_0}^{t_1}, \quad i = 1, 2, ..., n \tag{6.5}$$

using the usual definition of the canonical momenta $p_i = \frac{\partial L}{\partial \dot{q}_i}$. Now if both the actual and virtual paths coincide at $t_0$ and $t_1$, which are initial and final times, respectively, the virtual displacements $\delta q_i$ vanish at $t_0$ and $t_1$ implying from (6.5)

$$\delta S = 0 \tag{6.6}$$

We are thus in a position to state the principle of stationary action which says that the actual path chosen by a physical system, between end points $P_0(q_0, t_0)$ and $P_1(q_1, t_1)$, is such that along it the action (6.4) is stationary as

[1]Note that the action has the dimensions of energy × time or force × length × time.

compared with neighboring virtual paths (i.e., which are geometrically possible) having the same terminal points (namely, $P_0$ and $P_1$) as the actual trajectory. In other words, the time integral of the Lagrangian is an extremum: (6.6) is also called Hamilton's action principle. It was first proposed by Hamilton in 1834–35 and later extended to nonstationary constraints by Ostrogradsky in 1848. The principle of stationary action owes its formulation to Pierre Louis Maupertuis who proposed it in 1744 in analogy to Fermat's Principle of Least Time and Leonhard Euler who established Maupertuis's idea in the same year.

We have thus far exploited Lagrange's equations of motion to arrive at Hamilton's principle of stationary action (6.6). The converse also works. Since

$$\delta S = \int_{t_0}^{t_1} \delta L dt = \int_{t_0}^{t_1} \sum_{i=1}^{n} \left[ \frac{\partial L}{\partial q_i} - \frac{d}{dt}\left( \frac{\partial L}{\partial \dot{q}_i} \right) \right] \delta q_i dt$$

where $\delta q_i$ are arbitrary independent variations, $\delta S = 0$ provides Lagrange's equations of motion

$$\frac{d}{dt}\left( \frac{\partial L}{\partial \dot{q}_i} \right) - \frac{\partial L}{\partial q_i} = 0, \quad i = 1, 2, ..., n$$

by equating to zero the coefficient of each $\delta q_i$ (i=1,2,...,n) which are arbitrary and independent.

The principle of stationary action has an axiomatic status in classical mechanics. Lagrange's equations and the principle of stationary action are equivalent from an information content point of view. The action principle has its roots in the calculus of variations where the differential equations of the form as above are referred to as Euler's equations and emerge as a necessary condition for the existence of an extremum for the variational problem. The set of Lagrange's equations is also called Euler–Lagrange's equations.

**Proposition:**

The functionals $\int_{t_0}^{t_1} L(q, \dot{q}, t)dt$ and $\int_{t_0}^{t_1} \left[ L(q, \dot{q}, t) + \frac{dF}{dt} \right] dt$ lead to the same equations of motion.

**Proof:**

$$\begin{aligned} S' &= \int_{t_0}^{t_1} \left[ L(q, \dot{q}, t) + \frac{dF}{dt} \right] dt \\ &= \int_{t_0}^{t_1} L(q, \dot{q}, t)dt + [F(t_1) - F(t_0)] \\ \Rightarrow \delta S' &= \delta S + \delta \left[ F(t_1) - F(t_0) \right] \end{aligned}$$

It is obvious that the second term in the right-hand side vanishes. Hence by the principle of stationary action $\delta S' = \delta S = 0$ and we are led to a similar set of equations of motion from both $S$ and $S'$.

We conclude this section by making a few remarks on the passive and active points of view. In the passive point of view, if $C$ represents the path of an actual motion, then the observer in the unprimed reference frame will write his action as

$$S[C] = \int_{t_0}^{t_1} L(q, \dot{q}, t)dt$$

On the other hand, the observer in the primed reference frame will write his action as

$$S[C] = \int_{t_0'}^{t_1'} L'(q', \dot{q}', t')dt'$$

Since $dt'$ is not expected to be equal to $dt$, i.e., $\frac{dt'}{dt} \neq 1$, the functional form of $L$ and $L'$ would, in general, be different. However, the form of Lagrange's equations, obtained from the stationary character of the action, one for the unprimed system and one for the primed system, would be similar. This is known as covariance:

$$\begin{aligned} \frac{d}{dt}\left(\frac{\partial L}{\partial \dot{q}}\right) &= \frac{\partial L}{\partial q}, \\ \frac{d}{dt}\left(\frac{\partial L'}{\partial \bar{q}'}\right) &= \frac{\partial L'}{\partial \bar{q}'}, \qquad \bar{q}' \equiv \frac{dq'}{dt'} \end{aligned}$$

In the active point of view, for the actual path $C$, $S[C]$ is

$$S[C] = \int_{t_0}^{t_1} L(q, \dot{q}, t)dt$$

It is to be stationary in comparison to other neighboring paths. If we think of transformations which carry actual paths to actual paths, namely from $C$ to $C'$, then such transformations are called invariance transformations. Thus for invariance, the action for $C'$

$$S[C'] = \int_{t_0'}^{t_1'} L(q', \bar{q}', t')dt'$$

should be stationary as compared with neighboring paths to $C'$. Here $L$ remains the same. We shall deal with invariance transformation in some detail in Chapter 8 in connection with symmetries and conserved quantities of a physical system.

## 6.2 Corollaries

### (a) Hamilton's principle from D'Alembert's principle

The integral $\int_{t_0}^{t_1}(T-V)dt$ is stationary for an actual trajectory in comparison with neighboring paths having coordinates of the end points fixed along with the terminal time instants.

**Proof:**

We know from D'Alembert's principle that for a set of applied forces $\vec{F_i^a}$ the following holds:

$$\sum_{i=1}^{N}(\vec{F_i^a}-m_i\ddot{\vec{r_i}}).\delta\vec{r_i}=0$$

Since $\frac{d}{dt}(\delta\vec{r_i})=\delta\frac{d}{dt}(\vec{r_i})=\delta\dot{\vec{r_i}}$, it can also be expressed in the form

$$\sum_{i=1}^{N}m_i\left[\frac{d}{dt}(\dot{\vec{r_i}}.\delta\vec{r_i})-\dot{\vec{r_i}}.\delta\dot{\vec{r_i}}\right]=\sum_{i=1}^{N}\vec{F_i^a}.\delta\vec{r_i}$$

Identifying the second term in the left-hand side as the variation of the kinetic energy $T=\frac{1}{2}\sum_{j=1}^{N}m_j|\dot{\vec{r_j}}|^2$ and the right-hand side as the virtual work $\delta W$ for the applied forces we have the relation

$$\sum_{i=1}^{N}m_i[\frac{d}{dt}(\dot{\vec{r_i}}.\delta\vec{r_i})]=\delta T+\delta W \tag{6.7}$$

Integrating between $t_0$ and $t_1$ and noting that the coordinates of the end points are fixed there, the left-hand side clearly vanishes yielding

$$\int_{t_0}^{t_1}(\delta T+\delta W)dt=0 \tag{6.8}$$

The above result holds in general encompassing non-conservative systems as well.

It is to be stressed that while the virtual work $\delta W$ for the applied force is a well-defined quantity as given by the sum $\sum_{i=1}^{N}\vec{F_i^a}.\delta\vec{r_i}$, the same cannot be said of a finite $W$ which, in general, is a path-dependent entity. However, in the case of conservative systems where the existence of a potential function $V$ is assured we can express

$$\delta W=\sum_{i=1}^{N}\vec{F_i^a}.\delta\vec{r_i}=-\sum_{i=1}^{N}\frac{\partial V}{\partial\vec{r_i}}\cdot\delta\vec{r_i}=-\delta V \tag{6.9}$$

As a consequence, we recast $\int_{t_0}^{t_1}(\delta W)dt$ as -$\int_{t_0}^{t_1}(\delta V)dt$, i.e., -$\delta\int_{t_0}^{t_1} V dt$. We are therefore led to Hamilton's principle

$$\delta\int_{t_0}^{t_1}(T-V)dt = 0 \tag{6.10}$$

which in terms of the Lagrangian $L = T - V$ reads

$$\delta\int_{t_0}^{t_1} Ldt = 0 \tag{6.11}$$

signifying the extremum character of the time integral of the Lagrangian.

In the case of a free particle for which $V = 0$, we can express the action integral as simply $\int vdt = 0$ for a suitable time interval. This can be translated to $\int ds = 0$ which stands for the principle of shortest path.

**(b) Hamilton's canonical equations from the stationary action principle**

We can write

$$\delta\int_{t_0}^{t_1} Ldt = \int_{t_0}^{t_1}\delta Ldt = \int_{t_0}^{t_1}\sum_{i=1}^{N}\delta(p_i\dot{q}_i - H)dt$$

using the definition of the Hamiltonian. Taking the variation we can express the right-hand side as

$$\int_{t_0}^{t_1}\sum_{i=1}^{N}\left[p_i\delta\dot{q}_i + \left(\dot{q}_i - \frac{\partial H}{\partial p_i}\right)\delta p_i - \frac{\partial H}{\partial q_i}\delta q_i\right]dt$$

But $\delta\dot{q}_i = \frac{d}{dt}(\delta q_i)$ implying that we can write

$$\int_{t_0}^{t_1}(p_i\delta\dot{q}_i)dt = \int_{t_0}^{t_1} p_i\frac{d}{dt}(\delta q_i)dt = -\int_{t_0}^{t_1}\dot{p}_i.\delta q_i dt$$

because $\delta q_i$ vanishes at the terminal time points $t_0$ and $t_1$. Hence

$$\delta\int_{t_0}^{t_1} Ldt = \int_{t_0}^{t_1}\sum_{i=1}^{N}\left[-\left(\frac{\partial H}{\partial q_i} + \dot{p}_i\right)\delta q_i + \left(\dot{q}_i - \frac{\partial H}{\partial p_i}\right)\delta p_i\right]dt$$

The quantities $\delta q_i$ and $\delta p_i$ being arbitrary and independent, the principle of stationary action $\delta\int_{t_0}^{t_1} Ldt = 0$ gives Hamiltonian canonical equations:

$$\dot{q}_i = \frac{\partial H}{\partial p_i}, \qquad \dot{p}_i = -\frac{\partial H}{\partial q_i}$$

**Example 6.1**

Use Hamilton's principle to write down the equations of the motion in spherical polar coordinates.

In spherical polar coordinates the kinetic energy $T$ has the form

$$T = \frac{1}{2}\left(\dot{r}^2 + r^2\dot{\theta}^2 + r^2\sin^2\theta\dot{\phi}^2\right), \quad (m = 1 \text{ has been set })$$

Therefore, Hamilton's principle reads

$$\delta\int_{t_0}^{t_1}[\frac{1}{2}(\dot{r}^2 + r^2\dot{\theta}^2 + r^2\sin^2\theta\dot{\phi}^2) - V]dt = 0$$

where the potential $V$ is $V = V(r, \theta, \phi)$. Writing

$$\delta V = (\frac{\partial V}{\partial r}\delta r + \frac{\partial V}{\partial \theta}\delta\theta + \frac{\partial V}{\partial \phi}\delta\phi)$$

and noting that integrating by parts gives

$$\int_{t_0}^{t_1}\dot{r}\delta\dot{r}dt = -\int_{t_0}^{t_1}\ddot{r}\delta r dt$$

$$\int_{t_0}^{t_1} r^2\dot{\theta}\delta\dot{\theta}dt = -\int_{t_0}^{t_1}\frac{d}{dt}(r^2\dot{\theta})\delta\theta dt$$

$$\int_{t_0}^{t_1} r^2\sin^2\theta\dot{\phi}\frac{d}{dt}(\delta\phi)dt = -\int_{t_0}^{t_1}\frac{d}{dt}(r^2\sin^2\theta\dot{\phi})\delta\phi dt$$

where we have discarded terms evaluated at the boundaries $t_0$ and $t_1$ because of the fixed end points, we arrive at the expression

$$\int_{t_0}^{t_1}[(\ddot{r} - r\dot{\theta}^2 - r\sin^2\theta\dot{\phi}^2 + \frac{\partial V}{\partial r})\delta r + \{\frac{d}{dt}(r^2\dot{\theta}) - r^2\sin\theta\cos\theta\dot{\phi}^2 + \frac{\partial V}{\partial \theta}\}\delta\theta$$
$$+\{\frac{d}{dt}(r^2\sin^2\theta\dot{\phi}) + \frac{\partial V}{\partial \phi}\}\delta\phi]dt = 0$$

Since $\delta r, \delta\theta$ and $\delta\phi$ are arbitrary and independent variations we find on equating to zero the respective coefficients

$$\begin{aligned}
\ddot{r} - r\dot{\theta}^2 - r\sin^2\theta\dot{\phi}^2 &= -\frac{\partial V}{\partial r}, \\
\frac{1}{r}\frac{d}{dt}(r^2\dot{\theta}) - r\sin\theta\cos\theta\dot{\phi}^2 &= -\frac{1}{r}\frac{\partial V}{\partial \theta}, \\
\frac{1}{r\sin\theta}\frac{d}{dt}(r^2\sin^2\theta\dot{\phi}) &= -\frac{1}{r\sin\theta}\frac{\partial V}{\partial \phi}
\end{aligned}$$

which are the respective equations of motion for $r$, $\theta$ and $\phi$ variables.

## 6.3 Continuous systems: Uniform string problem

Thus far in this book we have handled systems with a finite number of generalized coordinates indexed by the label $i = 1, 2, ..., n$. In other words, we deal with systems with a finite number of degrees of freedom. However, in problems like that of a uniform string which constitute a continuous system we need to construct equations that are local and concern infinite degrees of freedom. Thus the generalized coordinates become continuous functions of space and time.

Suppose we have a tightly stretched uniform string under a tension $T$ which is fixed at both ends. Let the string have a length $l$ in its equilibrium position and mass $\mu$ per unit length. We take the $x$-axis along the length of the string. In equilibrium the string occupies the portion from $x = 0$ to $x = l$. If the string is set in vibration we can write for the kinetic energy at time $t$

$$\int_0^l \frac{1}{2}(\frac{\partial u}{\partial t})^2 \mu dx \tag{6.12}$$

where $u(x, t)$ denotes the vertical displacement of the string from its equilibrium position to the point $x$ at time $t$.

To find the potential energy, we note that for a slight displacement $\Delta x$ (that causes the length of the string, initially held tight, to increase by a little amount say, $\delta l$), work done against the force of tension $T$ is given by an element $dW$

$$dW = T \times (ds - dx) \tag{6.13}$$

where $ds$ is an element of the arc length given by

$$ds^2 \simeq dx^2 + du^2 \tag{6.14}$$

For small vibrations we can approximate

$$(\frac{ds}{dx})^2 = 1 + (\frac{\partial u(x,t)}{\partial x})^2 \tag{6.15}$$

enabling us to write

$$dW = T \times [\sqrt{1 + (\frac{\partial u(x,t)}{\partial x})^2} - 1]dx = \frac{1}{2}T(\frac{\partial u(x,t)}{\partial x})^2 dx \tag{6.16}$$

Thus the total potential energy which is the work done while the string is stretched to its displaced state from the equilibrium postion is

$$V = \int_0^l \frac{1}{2}T(\frac{\partial u(x,t)}{\partial x})^2 dx \tag{6.17}$$

The Lagrangian for the string can be written as

$$L = \int_0^l [\frac{1}{2}\mu(\frac{\partial u}{\partial t})^2 - \frac{1}{2}T(\frac{\partial u(x,t)}{\partial x})^2]dx \tag{6.18}$$

which can be reexpressed in terms of the Lagrangian density $\mathcal{L}$

$$L = \int_0^l \mathcal{L} dx \tag{6.19}$$

where

$$\mathcal{L} = \frac{1}{2}\mu(\frac{\partial u}{\partial t})^2 - \frac{1}{2}T(\frac{\partial u(x,t)}{\partial x})^2] \tag{6.20}$$

To derive the equations of motion for such a Lagrangian we can use Hamilton's principle by considering the variation $\delta u(x,t)$ on $u$ for fixed end points in time

$$\delta u(x,t_0) = \delta u(x,t_1) = 0 \tag{6.21}$$

The variations, of course, also vanish at the fixed end points of the string.

Turning to the action which is given by

$$S = \int_{t_0}^{t_1}\int_0^l \mathcal{L} dx \tag{6.22}$$

its variation is given by

$$\delta S = \int_{t_0}^{t_1}\int_0^l [\frac{\partial \mathcal{L}}{\partial \dot{u}}\delta\dot{u} + \frac{\partial \mathcal{L}}{\partial u'}\delta u' + \frac{\partial \mathcal{L}}{\partial u}\delta u]dx \tag{6.23}$$

where a dot and a prime denote, respectively, partial derivative with respect to $t$ and $x$. Interchanging as usual the operations of a variation and the partial derivative and integrating by parts with respect to $t$ in the first term and with respect to $x$ in the second term of the integrand leaves us with the result

$$\delta S = \int_{t_0}^{t_1}\int_0^l [-\frac{d}{dt}(\frac{\partial \mathcal{L}}{\partial \dot{u}}) - \frac{d}{dx}(\frac{\partial \mathcal{L}}{\partial u'}) + \frac{\partial \mathcal{L}}{\partial u}]dxdt\delta u(x,t) \tag{6.24}$$

The variations $u(x,t)$ being arbitrary and independent, for an action to be an extremum the integrand has to vanish giving the equations of motion

$$-\frac{d}{dt}(\frac{\partial \mathcal{L}}{\partial \dot{u}}) - \frac{d}{dx}(\frac{\partial \mathcal{L}}{\partial u'}) + \frac{\partial \mathcal{L}}{\partial u} = 0 \tag{6.25}$$

For the specific case of the Lagrangian of the string given above we find the form

$$\frac{\partial^2 u}{\partial t^2} = c^2\frac{\partial^2 u}{\partial x^2}, \quad c^2 = \frac{T}{\mu} \tag{6.26}$$

which gives the transverse equation of motion of a uniform string. Note that $c$ has the dimension of velocity and is the propagation velocity of the transverse waves. Equation (6.26) represents a one-dimensional linear wave equation.

To look for a general solution of (6.26) we invoke a change of variables specified by $\xi = x - ct, \quad \eta = x + ct$ which transform it to the form $u_{\xi\eta} = 0$. It transpires immediatey that its solution has a separable form consisting of arbitrary function of $\xi$ and $\eta$. Taking cue from it we can express the general solution of (6.26) in the form

$$u(x,t) = F(x - ct) + G(x + ct) \tag{6.27}$$

where $F$ and $G$ are arbitrary functions of their arguments.

Let us consider an initial value problem given by the initial conditions:

$$u(x,0) = f(x), \quad g_t(x,0) = g(x) \tag{6.28}$$

where $f$ and $g$ are given functions corresponding to initial displacement and initial velocity, respectively. Note that we are dealing with an evolution process and that (6.27), (6.28) together constitute the so-called Cauchy problem for the one-dimensional wave equation (6.26). The lines $x - ct =$ constant and $x+ct =$ constant are called the *characteristics* in the terminology of differential equations.

Given the above initial conditions we infer from (6.27) the following conditions that are needed to be satisfied:

$$F(x) + G(x) = f(x), \quad cF'(x) - cG'(x) = g(x) \tag{6.29}$$

where the prime represents a derivative. Solving for $F$ and $G$ we find

$$u(x,t) = \frac{1}{2}f(x + ct) + \frac{1}{2}g(x - ct) + \frac{1}{2c}\int_{x-ct}^{x+ct} g(s)ds \tag{6.30}$$

(6.30) is called D'Alembert's solution for the wave equation. It shows that the solution depends only upon the initial value functions specified at $x - ct$ and $x + ct$ and on the integral of $g$ evaluated between these two points. In the case of zero initial velocity we have a simple superposition

$$u(x,t) = \frac{1}{2}f(x + ct) + \frac{1}{2}g(x - ct) \tag{6.31}$$

showing two waves travelling to the right and left with the same speed $c$.

To show that D'Alembert's solution is well-posed we take a set of two solutions $u_1$ and $u_2$ corresponding, respectively, to two initial specifications of functions $f_1, f_2$ and $g_1, g_2$. Suppose that

$$|f_1(x) - f_2(x)| < \delta, \quad |g_1(x) - g_2(x)| < \delta \tag{6.32}$$

where $\delta$ is an arbitrary small preassigned number. Since from (6.30)

$$\begin{aligned} u_1 - u_2 &= \frac{1}{2}[f_1(x-ct) - f_2(x+ct) + g_1(x-ct) - g_2(x+ct)] \\ &\quad + \frac{1}{2c}\int_{x-ct}^{x+ct}[g_1(s) - g_2(s)]ds \end{aligned} \tag{6.33}$$

then for all $x \in \mathbb{R}$ and $0 \leq t \leq \tau$, it follows by (6.32)

$$|u_1 - u_2| < \frac{1}{2}(\delta + \delta) + \frac{1}{2c}(x + ct - x + ct) < \delta(1+\tau) < \epsilon \tag{6.34}$$

where $\delta$ has been taken $< \frac{\epsilon}{1+\tau}$ for a given $\epsilon > 0$. We therefore conclude that small changes in the initial conditions produce a corresponding small change in the solutions implying that the Cauchy problem is well-posed.

---

## 6.4 Normal modes of oscillation

To find normal modes of oscillations, we note that the string equation is subject to the boundary conditions

$$u(0,t) = 0, \quad u(l,t) = 0 \tag{6.35}$$

If we seek a solution of the form

$$u(x,t) = \psi(x)\cos(\omega t + \phi) \tag{6.36}$$

which is in a separable form involving an arbitrary function $\psi(x)$ and a trigonometric factor of time chosen due to the linearity of the differential equation, then substitution in (6.30) shows that $\psi$ satisfies the differential equation

$$\frac{d^2\psi}{dt^2} + \frac{\omega^2}{c^2}\psi = 0 \tag{6.37}$$

The accompanying boundary conditions as guided by (6.34) are

$$\psi(0) = 0, \quad \psi(l) = 0 \tag{6.38}$$

In the general solution of (6.37) which is of the form

$$\psi(x) = A\cos\frac{\omega x}{c} + B\sin\frac{\omega x}{c} \tag{6.39}$$

where $A$ and $B$ are arbitrary constants, $A$ turns out to be zero while $\omega$ is restricted by the constraint $l\omega = nc\pi$, where $n$ is an integer. As a result the $n^{th}$ normal mode of the string is given by

$$u_n(x,t) = B_n \sin\frac{n\pi x}{l}\cos(\omega_n + \phi) \tag{6.40}$$

Hence $u(x,t)$ is expressible in the form

$$u(x,t) = \sum_{n=1}^{\infty} \sin\frac{n\pi x}{l}[C_n \cos\frac{n\pi ct}{l} + D_n \sin\frac{n\pi ct}{l}] \tag{6.41}$$

where $C_n$ and $D_n$ are arbitrary constants and can be fixed from the initial conditions.

To this end we make use of the first condition in (6.28) to write down

$$u(x,0) = f(x) = \sum_{n=1}^{\infty} C_n \sin\frac{n\pi x}{l} \tag{6.42}$$

which on inversion gives

$$C_n = \frac{2}{l}\int_0^l f(x)\sin\frac{n\pi x}{l}dx \tag{6.43}$$

On the other hand, if we differentiate with respect to $t$ and use the second of (6.28) then

$$u_t(x,0) = g(x) = \sum_{n=1}^{\infty} D_n \frac{n\pi c}{l}\sin\frac{n\pi x}{l} \tag{6.44}$$

On inverting this gives for $D_n$

$$D_n = \frac{2}{n\pi c}\int_0^l g(x)\sin\frac{n\pi x}{l}dx \tag{6.45}$$

We wish to remark that in the case of zero initial velocity the solution (6.41) because of (6.45) reduces to

$$u(x,t) = \sum_{n=1}^{\infty} C_n \sin\frac{n\pi x}{l}\cos\frac{n\pi ct}{l} \tag{6.46}$$

where $C_n$ is given by (6.43). Making use of the trigonometric identity $\sin A\cos B = \frac{1}{2}[\sin(A-B)+\sin(A+B)]$, we deduce from (6.46) the expression for $u(x,t)$

$$u(x,t) = \sum_{n=1}^{\infty} C_n[\sin\frac{n\pi(x-ct)}{l} + \sin\frac{n\pi(x+ct)}{l}] \tag{6.47}$$

which can be recognized to be in D'Alembert's form (6.31).

## 6.5 Extended point transformation and $\Delta$ variation

There are certain physical situations in which moving boundaries are relevant. This is in contrast to the case of Hamilton's principle where the varied path shares with the actual path the same end points which are fixed.

Curves with variable boundary points can appear as a result of extended point mappings transforming $(q,t)$ to the new forms $(q',t')$ as given by $q' = q'(q,t), t' = t'(q,t)$. We restrict ourselves to the infinitesimal case

$$q' = q + \Delta q(q,t), \qquad t' = t + \Delta t(q,t) \tag{6.48}$$

where $\Delta q$ and $\Delta t$ are infinitesimal small changes. Note that $\Delta$ is to be distinguished from $\delta$ in that we use $\delta$ for a virtual (time-frozen) change.

In the active point of view there is a single observer who observes the evolution of the system paths $C$ to $C'$. To him the difference in the action for two neighboring paths appears as

$$\Delta S = S[C'] - S[C] = \int_{t_0}^{t_1} \left[ L\left(q', \frac{dq'}{dt'}, t'\right) dt' - L\left(q, \frac{dq}{dt}, t\right)\right] dt \tag{6.49}$$

where $(q',t')$ are infinitesimally different from $(q,t)$.

We can reexpress (6.48) as

$$dq' = dq + \left(\frac{d\Delta q}{dt}\right) dt, \quad dt' = dt + \left(\frac{d\Delta t}{dt}\right) dt \tag{6.50}$$

where $\frac{d}{dt} = \frac{\partial}{\partial t} + \dot{q}\frac{\partial}{\partial q}$ as usual. Since $\Delta q$ and $\Delta t$ are both infinitesimal we have for the ratio

$$\frac{dq'}{dt'} = \left(1 - \frac{d\Delta t}{dt}\right)\frac{dq}{dt} + \frac{d\Delta q}{dt} \tag{6.51}$$

yielding the non-trivial result

$$\frac{dq'}{dt'} - \frac{dq}{dt} \neq \frac{d\Delta q}{dt}$$

i.e., change in the generalized velocities is not the same as the time derivative of the change in the generalized coordinate.

Using (6.50) and (6.51), $L\left(q', \frac{dq'}{dt'}, t'\right)$ can be expanded as

$$\begin{aligned}
L\left(q', \frac{dq'}{dt'}, t'\right) dt' &= L\left[q + \Delta q, \left(1 - \frac{d\Delta t}{dt}\right)\frac{dq}{dt} + \frac{d\Delta q}{dt}, t + \Delta t\right]\left(1 + \frac{d\Delta t}{dt}\right) \\
&= L\left[q + \Delta q, \frac{dq}{dt} + \{\frac{d\Delta q}{dt} - \left(\frac{d\Delta t}{dt}\right)\frac{dq}{dt}\}, t + \Delta t\right]\left(1 + \frac{d\Delta t}{dt}\right) \\
&= \left[L\left(q, \frac{dq}{dt}, t\right) + \Delta q\frac{\partial L}{\partial q} + \{\frac{d\Delta q}{dt} - \left(\frac{d\Delta t}{dt}\right)\frac{dq}{dt}\}\frac{\partial L}{\partial \dot{q}} + \Delta t\frac{\partial L}{\partial t}\right]\left(1 + \frac{d\Delta t}{dt}\right) \\
&= L\left(q, \frac{dq}{dt}, t\right) + \frac{\partial L}{\partial q}\Delta q + \frac{\partial L}{\partial \dot{q}}\left(\frac{d\Delta q}{dt} - \dot{q}\frac{d\Delta t}{dt}\right) + \frac{\partial L}{\partial t}\Delta t + L\frac{d\Delta t}{dt}
\end{aligned}$$

where we have kept only the first order quantities in $\Delta q, \Delta t$ and used Taylor expansion

$$\phi(x+\Delta x, y+\Delta y, z+\Delta z, ...) = \phi(x,y,z,...) + \left(\Delta x\frac{\partial}{\partial x} + \Delta y\frac{\partial}{\partial y} + \Delta z\frac{\partial}{\partial z} + ...\right)\phi(x,y,z,...) + ...$$

Thus (6.49) becomes

$$\Delta S = \int_{t_0}^{t_1}\left[\frac{\partial L}{\partial q}\Delta q + \frac{\partial L}{\partial \dot q}\left(\frac{d\Delta q}{dt} - \dot q\frac{d\Delta t}{dt}\right) + \frac{\partial L}{\partial t}\Delta t + L\frac{d\Delta t}{dt}\right]dt$$

The above expression can be simplified further because by integration by parts we can write

$$\int \frac{\partial L}{\partial \dot q}\frac{d\Delta q}{dt}dt = \frac{\partial L}{\partial \dot q}\Delta q - \int \frac{d}{dt}\left(\frac{\partial L}{\partial \dot q}\right)\Delta q dt$$

$$\int \frac{\partial L}{\partial \dot q}\dot q\frac{d\Delta t}{dt}dt = \frac{\partial L}{\partial \dot q}\dot q\Delta t - \int \frac{d}{dt}\left(\frac{\partial L}{\partial \dot q}\dot q\right)\Delta t dt$$

$$\int L\frac{d\Delta t}{dt}dt = L\Delta t - \int \frac{dL}{dt}\Delta t dt$$

which yields the form

$$\Delta S = \int_{t_0}^{t_1} dt\left[(\Delta q - \dot q\Delta t)\{\frac{\partial L}{\partial q} - \frac{d}{dt}(\frac{\partial L}{\partial \dot q})\}\right] + \left[\frac{\partial L}{\partial \dot q}\Delta q + \left(L - \frac{\partial L}{\partial \dot q}\dot q\right)\Delta t\right]_{t_0}^{t_1}$$

where we have expressed the difference $\frac{dL}{dt} - \frac{\partial L}{\partial t} = \dot q\frac{\partial L}{\partial q}$. For the actual path Lagrange's equations are satisfied and hence we are left only with

$$\Delta S = \left[\frac{\partial L}{\partial \dot q}\Delta q + \left(L - \frac{\partial L}{\partial \dot q}\dot q\right)\Delta t\right]_{t_0}^{t_1} \tag{6.52}$$

An interesting special case of (6.52) is that when the the system is conservative and scleronomic having its kinetic energy as a homogeneous function of velocities. Corresponding to the Hamiltonian $H$ and the canonical momentum $p$ we then have

$$\Delta S = [p\Delta q - H\Delta t]_{t_0}^{t_1} \tag{6.53}$$

For $N$ number of particles, (6.53) can be extended to

$$\Delta S \equiv \Delta\int_{t_0}^{t_1} Ldt = \sum_{i=1}^{N}[p_i\Delta q_i - H\Delta t]_{t_0}^{t_1} \tag{6.54}$$

If we restrict to the following prescriptions:

(i) $H$ does not depend explicitly on $t$ and so $H$ is conserved.
(ii) $H$ is conserved not only on the actual path but also on the varied path and
(iii) Varied paths are constrained such that $\Delta q_i (i = 1, 2, ..., n)$ vanish at the end points but not $\Delta t$.
then (6.54) further simplifies to

$$\Delta S \equiv \Delta \int_{t_0}^{t_1} L dt = -H(\Delta t_1 - \Delta t_0) \tag{6.55}$$

Now, the action integral reads

$$\int_{t_0}^{t_1} L dt = \int_{t_0}^{t_1} \left[ \sum_{i=1}^{n} p_i \dot{q}_i - H \right] dt = \int_{t_0}^{t_1} \sum_{i=1}^{n} p_i \dot{q}_i dt - H(t_1 - t_0) \tag{6.56}$$

Taking $\Delta$ variation of both sides we have

$$\Delta \int_{t_0}^{t_1} L dt = \Delta \int_{t_0}^{t_1} \sum_{i=1}^{n} p_i \dot{q}_i dt - H(\Delta t_1 - \Delta t_0) \tag{6.57}$$

If we compare with (6.55) we find

$$\Delta \int_{t_0}^{t_1} \sum_{i=1}^{n} p_i \dot{q}_i dt = 0 \tag{6.58}$$

The integral in (6.58) is generally referred in old books as the action or the action integral and (6.58) as the principle of least action. Historically it was Fermat who made an application of it in the case of refraction of light. In recent times it is more customary to refer to the integral in Hamilton's principle as the action. Recall Hamilton's principle is the variational principle which states that the physical trajectory is the one for which the action is stationary.

If $L = T - V$ and $H = T + V$ then

$$\begin{aligned} \sum_{i=1}^{n} p_i \dot{q}_i &= H + L \\ &= T + V + (T - V) \\ &= 2T \end{aligned}$$

Hence another form of (6.58) is

$$\Delta \int_{t_0}^{t_1} 2T dt = 0 \tag{6.59}$$

## 6.6 $\Delta$ and $\delta$ variations

For the terminal time points $t_0$ and $t_1$ we already know that $\delta(t_0) = \delta(t_1) = 0$. Consider a function $f(t)$. We can expand it as

$$f_1(t + \Delta t) - f(t) = f_1(t) + \Delta t \dot{f}_1 - f(t) = [f_1(t) - f(t)] + \Delta t \dot{f}_1 = \delta f + \Delta t \dot{f}_1$$

Writing $f_1 = f + \Delta f$ we obtain

$$\Delta f = \delta f + \Delta t \dot{f} \tag{6.60}$$

where obviously $\Delta f = \delta f$ if $\Delta t = 0$. Differentiating (6.60) with respect to $t$ we have

$$\begin{aligned} \frac{d}{dt}(\Delta f) &= \frac{d}{dt}(\delta f) + \frac{d}{dt}(\Delta t)\dot{f} + \Delta t \frac{d}{dt}(\dot{f}) \\ &= \left[\delta \dot{f} + \Delta t \frac{d}{dt}(\dot{f})\right] + \frac{d}{dt}(\Delta t)\dot{f} \\ &= \Delta \dot{f} + \frac{d}{dt}(\Delta t)\dot{f} \end{aligned} \tag{6.61}$$

where we have used $\Delta \dot{f} = \delta \dot{f} + \Delta t \ddot{f}$ analogous to (6.60).

We thus arrive at the general formula

$$\frac{d}{dt}(\Delta f) = \Delta \dot{f} + \frac{d}{dt}(\Delta t)\dot{f} \tag{6.62}$$

showing that $\frac{d}{dt}$ and $\Delta$ are not interchangeable: $\frac{d}{dt}(\Delta) \neq \Delta \frac{d}{dt}$.

### Parametric representation of (6.59)

We now proceed to express (6.59) as a parametric equation. Toward this end writing $A = \int_{t_0}^{t_1} 2T dt$, we have

$$\begin{aligned} A &= \int_{t_0}^{t_1} 2T dt \\ &= \int_{t_0}^{t_1} \sqrt{2T}\sqrt{2T} dt \\ &= \int_{t_0}^{t_1} \sqrt{2(H-V)}\sqrt{\sum m(\frac{ds}{dt})^2} dt \\ &= \int_{t_0}^{t_1} \sqrt{2(H-V)}\sqrt{\sum m ds^2} \\ &= \int_{P_0}^{P_1} \sqrt{2(H-V)}\sqrt{\sum m(\frac{ds}{d\lambda})^2} d\lambda \end{aligned}$$

where $\lambda$ is an arbitrary parameter.

Thus $A$ can be written in an equivalent form

$$A = \int_{P_0}^{P_1} I d\lambda \tag{6.63}$$

where $I$ is given by $I = \sqrt{2(H-V)}\sqrt{\sum m(\frac{ds}{d\lambda})^2}$. Thus $I$ may be looked upon as a function of the generalized coordinates $(q_1, q_2, ..., q_N)$ and the derivatives $(\frac{dq_1}{d\lambda}, \frac{dq_2}{d\lambda}, ..., \frac{dq_N}{d\lambda})$ denoted by $(q_1', q_2', ..., q_N')$ where $q_i' \equiv \frac{dq_i}{d\lambda}, i = 1, 2, ..., N$.

**Proposition:**

The quantity $I$ obeys the differential equations

$$\frac{d}{d\lambda}\left(\frac{\partial I}{\partial \dot{q}_i}\right) = \frac{\partial I}{\partial q_i} \tag{6.64}$$

where $i = 1, 2, ..., N$.

**Proof:**

From (6.59), i.e., $\Delta A = 0$, we have on using (6.60) and (6.63)

$$\begin{aligned} 0 &= \delta \int_{P_0}^{P_1} I d\lambda \qquad \text{(as } I \text{ is independent of } t) \\ &= \int_{P_0}^{P_1} \sum_{r=1}^{n} \left[\frac{\partial I}{\partial q_r}\delta q_r + \frac{\partial I}{\partial q_r'}\delta q_r'\right] d\lambda \end{aligned} \tag{6.65}$$

Now since we can express

$$\begin{aligned} \int_{P_0}^{P_1} \frac{\partial I}{\partial q_r'}\delta q_r' d\lambda &= \int_{P_0}^{P_1} \frac{\partial I}{\partial q_r'}\frac{d}{d\lambda}(\delta q_r) d\lambda \\ &= \left[\frac{\partial I}{\partial q_r'}\delta q_r\right]_{P_0}^{P_1} - \int_{P_0}^{P_1} \frac{d}{d\lambda}(\frac{\partial I}{\partial q_r'})\delta q_r d\lambda \\ &= -\int_{P_0}^{P_1} \frac{d}{d\lambda}(\frac{\partial I}{\partial q_r'})\delta q_r d\lambda \qquad \text{(as } \delta q_r = 0 \text{ at } P_0 \text{ and } P_1) \end{aligned}$$

substituting it in (6.65) we get

$$\int_{P_0}^{P_1} \sum_{r=1}^{n} \left[\frac{\partial I}{\partial q_r} - \frac{d}{d\lambda}(\frac{\partial I}{\partial q_r'})\right] \delta q_r d\lambda = 0$$

Since the variations $\delta q_r$ are arbitrary and independent, the proposition follows.

**Example 6.2:**

A particle of unit mass is projected so that its total energy is $E$ in a field of force whose potential is $\phi(r)$ at distance $r$ from the origin. Deduce the differential equation of the path to be

$$C^2\left[r^2 + (\frac{dr}{d\theta})^2\right] = r^4[h - \phi(r)]$$

where $C$ is a constant.

Here $H = E$, $V = \phi(r)$ and in the $(r, \theta)$ plane $ds^2 = dr^2 + r^2 d\theta^2$. Therefore

$$\begin{aligned} A &= \int_{t_0}^{t_1} \sqrt{2T}\sqrt{2T}dt \\ &= \int_{t_0}^{t_1} \sqrt{2(H-V)}\sqrt{\sum m(\frac{ds}{dt})^2 dt} \\ &= \int_{t_0}^{t_1} \sqrt{2(E-\phi)}\sqrt{\sum m ds^2} \\ &= \int_{P_0}^{P_1} \sqrt{2(E-\phi)}\sqrt{\sum m(\frac{ds}{d\lambda})^2 d\lambda} \\ &= \int_{P_0}^{P_1} \sqrt{2(E-\phi)}\sqrt{(\frac{dr}{d\lambda})^2 + r^2(\frac{d\theta}{d\lambda})^2 d\lambda} \end{aligned}$$

where $\lambda$ is a parameter other than $t$. Therefore $I$ can be identified to be

$$I = \sqrt{2(E-\phi)}\sqrt{r'^2 + r^2\theta'^2}$$

where $r' \equiv \frac{dr}{d\lambda}, \theta' \equiv \frac{d\theta}{d\lambda}$. Since $\theta$ is an ignorable coordinate we deduce $\frac{d}{d\lambda}(\frac{\partial I}{\partial \theta'}) = 0$ from (6.64). As a result $\frac{\partial I}{\partial \theta'} =$ constant. Let the constant be $\sqrt{2}C$ implying

$$\sqrt{2(E-\phi)}\frac{r^2\theta'}{r'^2 + r^2\theta'^2} = \sqrt{2}C$$

which yields on some rearrangements the desired expression

$$C^2\left[r^2 + (\frac{dr}{d\theta})^2\right] = r^4[E - \phi(r)]$$

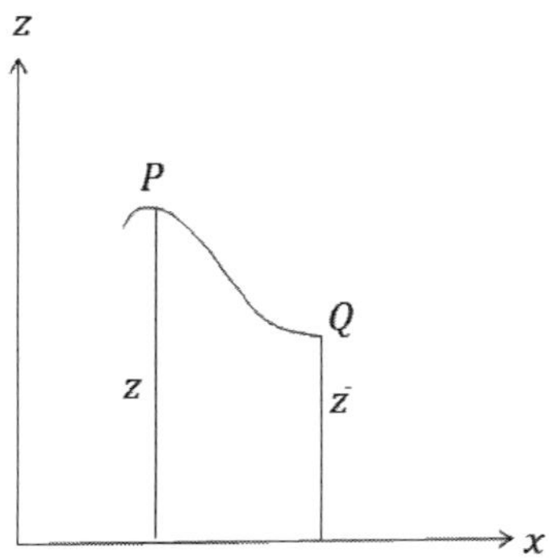


**FIGURE 6.2**: Brachistochrone curve.

## 6.7 Brachistochrone problem

The brachistochrone problem is due to Johann Bernoulli who in 1696 attempted to find a plane curve that joins two points such that the time required for a particle travelling under the influence of gravity takes the minimum time to move from one point to the other. The problem was subsequently solved among others by Leibniz, L'Hospital and Newton.

In Figure 6.2 we take the $x$-axis to be horizontal and $z$-axis as vertically upward. We consider a small element of *arc* $PQ$ corresponding to the initial and final positions $P$ and $Q$ of the particle, respectively. If the velocity and height of the particle at P be $(v, z)$ and the same at Q be $(\bar{v}, \bar{z})$ then it is clear that

$$\begin{aligned} \text{total energy at P} &= \frac{1}{2}mv^2 + mgz \\ \text{total energy at Q} &= \frac{1}{2}m\bar{v}^2 + mg\bar{z} \end{aligned}$$

Conservation of energy requires these to be equal so that

$$v^2 = \bar{v}^2 + 2g(\bar{z} - z) \tag{6.66}$$

Now we can express $v = \frac{ds}{dt}$ where $ds$ stands for the element PQ and $dt$ specifies the short interval of time. Hence the above equation can be expressed as

$$\begin{aligned} t &= \int_P^Q ds\left[\bar{v}^2 + 2g(\bar{z} - z)\right]^{-\frac{1}{2}} \\ &= \int_z^{z_0} \frac{\left[1 + (\frac{dx}{dz})^2\right]^{\frac{1}{2}}}{\left[\bar{v}^2 + 2g(\bar{z} - z)\right]^{\frac{1}{2}}} dz \end{aligned} \tag{6.67}$$

where we have put $ds = \sqrt{dx^2 + dz^2}$.

The integral above has the same form as that of an action with $x$ playing the role of a "generalized coordinate" and $z$ that of "time." The integrand defines the following Lagrangian:

$$\begin{aligned} L &= \frac{\left[1+(\frac{dx}{dz})^2\right]^{\frac{1}{2}}}{\left[\bar{v}^2+2g(\bar{z}-z)\right]^{\frac{1}{2}}} \\ &= \frac{\left(1+x'^2\right)^{\frac{1}{2}}}{\left[\bar{v}^2+2g(\bar{z}-z)\right]^{\frac{1}{2}}}, \qquad x'=\frac{dx}{dz} \\ &= L(x,x',z) \end{aligned} \tag{6.68}$$

Note $x$ is an ignorable coordinate here.

The path of minimum time is obtainable by taking an extremum of the integral

$$\delta \int_z^{z_0} L(x,x',z)dz = 0 \tag{6.69}$$

which yields

$$\frac{d}{dz}(\frac{\partial L}{\partial x'}) = \frac{\partial L}{\partial x} = 0 \tag{6.70}$$

From this we deduce

$$\begin{aligned} \frac{\partial L}{\partial x'} &= \text{constant} \\ \text{i.e.,} \quad \frac{x'}{(1+x'^2)^{\frac{1}{2}}\left[\bar{v}^2+2g(\bar{z}-z)\right]^{\frac{1}{2}}} &= C \text{ (say)} \\ \text{or} \quad x'^2(1+x'^2)^{-1}\left[\bar{v}^2+2g(\bar{z}-z)\right]^{-1} &= C^2 \end{aligned} \tag{6.71}$$

which gives

$$\begin{aligned} \frac{1}{x'^2} &= \frac{1-C^2\left[\bar{v}^2+2g(\bar{z}-z)\right]}{C^2\left[\bar{v}^2+2g(\bar{z}-z)\right]} \\ \text{or,} \quad (\frac{dx}{dz})^2 &= \frac{b-z}{a+z} \end{aligned} \tag{6.72}$$

where the constants $a$ and $b$ are

$$a = \frac{1-C^2\bar{v}^2-2gC^2\bar{z}}{2gC^2}, \qquad b = \frac{(\bar{v}^2+2g\bar{z})}{2g} \tag{6.73}$$

Setting

$$z = -(\frac{a-b}{2}) - (\frac{a+b}{2})\cos\theta \tag{6.74}$$

facilitates integration since we get the form

$$dx = (\frac{a+b}{2})\int\sqrt{\frac{1+\cos\theta}{1-\cos\theta}}\sin\theta d\theta \tag{6.75}$$

which implies

$$x = \frac{a+b}{2}(\theta - \sin\theta) + \text{ constant} \tag{6.76}$$

The variables $z$ and $x$ provide the curve for minimum time, the *brachistochrone*, in parametric form with both $x$ and $z$ appearing as functions of $\theta$. The curve can be easily recognized to be a cycloid.

---

## 6.8 Summary

We gave a general treatment of an action principle by taking recourse to variations which puts it in an elegant integral representation with the Lagrangian serving as the integrand between two terminal points of time. What the action principle says is that of all the possible paths, as the system traces out a trajectory from a given initial time $t_0$ to a final time $t_1$, the actual path, which is a solution of the equation of motion, is the one for which the action is an extremum. Interesting corollaries that follow from the action principle are the Lagrangian equations of motion and the Hamilton's canonical equations of motion. We considered the extended point transformations which are relevant for moving boundary problems and derived a particular variant of action principle. We also addressed briefly the brachistochrone problem.

# Exercises

1. Consider the integral

$$\int_{t_0}^{t_1} f(x)\sqrt{1+\dot{x}^2}dt \tag{6.77}$$

involving a function $f(x)$. Show that the extremum of the integral leads to the differential equation for $f(x)$

$$1 + \dot{x}^2 = cf(x)^2 \tag{6.78}$$

where $c$ is a constant.

2. A particle oscillates in a straight line about a center of force which varies as the distance. By integrating from $t_0$ to $t$ deduce the action in the form

$$\frac{\sqrt{\mu}}{2}\frac{(x_0^2 + x^2)\cos\sqrt{\mu}(t-t_0) - 2xx_0}{\sin\mu(t-t_0)} \tag{6.79}$$

where $\mu$ is the constant of proportionality.

3. Consider a particle of unit mass moving in the $xy$ plane under a central acceleration proportional to the radial distance. By integrating from 0 to $t$ determine the action to be

$$\frac{n}{2\sin nt}[(x_0^2 + x^2 + y_0^2 + y^2)\cos nt - 2(xx_0 + yy_0)] \tag{6.80}$$

where $n^2$ is the constant of proportionality.

4. Show that if time is coordinated as an additional coordinate of a mechanical system, then the momentum associated with time is the Hamiltonian of the system with its sign changed. Deduce that the total energy of a scleronomic system is a constant of motion.

5. Consider the integral

$$I = \int_{t_0}^{t_1} f dt \tag{6.81}$$

for the motion along an actual path and the integral

$$I_1 = \int_{t_0+\Delta t_0}^{t_1+\Delta t_1} f_1 dt \tag{6.82}$$

for a varied path. Show that the difference $\Delta I = I_1 - I$ is given by the integral representation

$$\Delta I = \int_{t_0}^{t_1} [\delta f + \frac{d}{dt}(f\Delta t)]dt \tag{6.83}$$

# Chapter 7

## Motion in noninertial coordinate systems



In Chapter 1 we introduced the idea of an inertial frame and noted that any coordinate frame moving uniformly with respect to it is inertial. We also encountered Galilean transformations that leave invariant the force-free equations of motion as derived from Newton's second law. In fact, use of the second law is limited only to inertial frames. However at certain times it is necessary for certain physical problems, to deal with non-inertial reference frames as in the case of motion occurring near the surface of the rotating Earth. Here introduction of a non-inertial frame simplifies the situation. Non-inertial frames arise, in general, for rotating frames and accelerated coordinate systems. The aim in this chapter is to study particle dynamics in a rotating frame of coordinates and provide a detailed derivation of forces that need be considered along with the external forces that exert on the system. The time independent harmonic restoring force, namely, the centrifugal force and the time-dependent Coriolis force, are two typical forces which appear in rotating reference frames. In particular, the role of the latter is to rotate the velocity vector making it precess around the rotation vector direction.

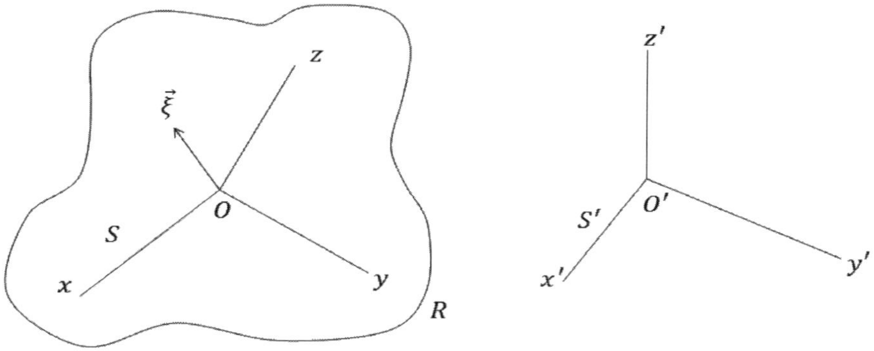


**FIGURE 7.1**: Rotating coordinate system $S$ fixed in the rigid body R and inertial coordinate system $S'$ fixed in space.

## 7.1 Rotating frames

### 7.1.1 Basic equations

A rotating frame of reference is a special case of a non-inertial system. Because of rotation it generates pseudo-forces as we shall presently see. Consider two orthogonal coordinate frames of reference namely S:OXYZ attached to a rigid body $R$ with standard unit basis vectors $i, j, k$ and S':O'X'Y'Z' a fixed one in space representing an inertial coordinate system (see Figure 7.1).

If the rigid body is rotating with an angular velocity $\vec{\omega}$ about a fixed axis through $O$, then it is obvious that for a vector $\vec{\xi}$ fixed in $R$, an observer positioned at $O$ will see no change in the components $(\xi_x, \xi_y, \xi_z)$ of $\vec{\xi}$ relative to $OX, OY, OZ$. However to an inertial observer at $O'$, the time-rate of change of $\vec{\xi}$ will appear as

$$\frac{d\vec{\xi}}{dt} = \left(\frac{d\xi_x}{dt}\vec{i} + \frac{d\xi_y}{dt}\vec{j} + \frac{d\xi_z}{dt}\vec{k}\right) + \xi_x\frac{d\vec{i}}{dt} + \xi_y\frac{d\vec{j}}{dt} + \xi_z\frac{d\vec{k}}{dt} \tag{7.1}$$

where $\frac{d\vec{i}}{dt} = \vec{\omega} \times \vec{i}$, $\frac{d\vec{j}}{dt} = \vec{\omega} \times \vec{j}$, $\frac{d\vec{k}}{dt} = \vec{\omega} \times \vec{k}$ are to be interpreted as induced velocities caused by the angular velocity of the frame $S$ relative to $S'$.

It therefore follows that

$$\left(\frac{d}{dt}\right)_{\text{fixed}} \vec{\xi} = \left(\frac{d}{dt}\right)_{\text{rot}} \vec{\xi} + \vec{\omega} \times \vec{\xi} \tag{7.2}$$

where $\left(\frac{d}{dt}\right)_{\text{rot}}$ stands for the time-rate as measured from the rotating frame $S$. Indeed Equation (7.2) furnishes the typical motion of a rigid body being described by a combination of translation and rotation. The suffixes "fixed"

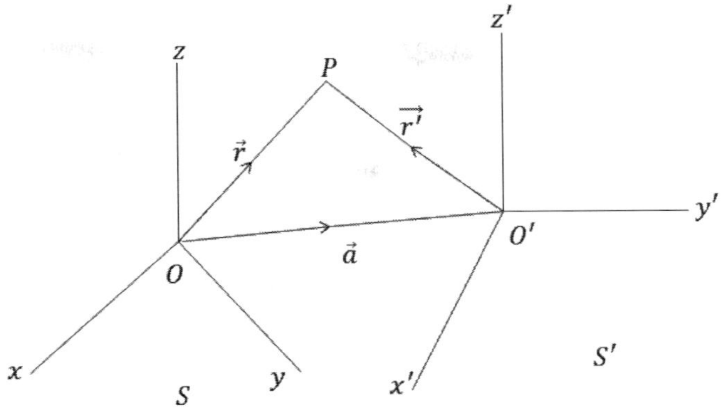


**FIGURE 7.2**: Position of a particle at $P$ viewed from two coordinate systems $S$ and $S'$.

and "rot" specify the roles of a fixed observer in $S'$ and someone rotating with the rigid body.

To derive the governing equations of a motion of a particle moving relative to a rotating frame, let us imagine that the coordinate frame $S'$ has been set up in a fixed star and the rigid body is our Earth itself. We neglect the orbital motion of Earth around the Sun and assume the Earth to be a perfect sphere.

Let $P$ be the position of a particle at time $t$ with $\vec{OP} = \vec{r}$ and $\vec{O'P} = \vec{r'}$. If $\vec{O'O} = \vec{a}$ then it is clear from Figure 7.2 that $\vec{r'} = \vec{r} + \vec{a}$. As such the velocity of the particle at $P$ when measured from $O'$ would read

$$\vec{u} \equiv \left(\frac{d}{dt}\right)_{\text{fixed}} \vec{r'} = \left(\frac{d}{dt}\right)_{\text{fixed}} \vec{r} + \frac{d\vec{a}}{dt} \tag{7.3}$$

To estimate $\left(\frac{d}{dt}\right)_{\text{fixed}} \vec{r}$ we substitute the relationship (7.2) between the fixed and a rotating frame replacing $\vec{\xi}$ by $\vec{r}$. In consequence we get from (7.3)

$$\vec{u} \equiv \left(\frac{d\vec{r}}{dt}\right)_{\text{rot}} + \vec{\omega} \times \vec{r} + \frac{d\vec{a}}{dt} \tag{7.4}$$

The first term in the right-hand side of (7.4) gives the velocity of the particle relative to the rotating frame $S$, the second term appears as a result of the rotation of the $OXYZ$ frame and the third term represents the so-called drag velocity. The drag velocity can be ignored if the distance vector between the points $O$ and $O'$ does not change with time.

To derive an expression for the acceleration of the particle at $P$ as measured from $O'$, we operate upon (7.4) by the derivative $\left(\frac{d}{dt}\right)_{\text{fixed}}$:

$$\left(\frac{d}{dt}\right)_{\text{fixed}} \vec{u} \equiv \left(\frac{d^2}{dt^2}\right)_{\text{fixed}} \vec{r'} = \left(\frac{d}{dt}\right)_{\text{fixed}} \left(\frac{d\vec{r}}{dt}\right)_{\text{rot}} + \left(\frac{d}{dt}\right)_{\text{fixed}} (\vec{\omega}\times\vec{r}) + \frac{d^2\vec{a}}{dt^2} \tag{7.5}$$

We now make use of the following results obtained by replacing $\vec{\xi}$ successively by the vectors $\left(\frac{d\vec{r}}{dt}\right)_{\text{rot}}$ and $\vec{\omega} \times \vec{r}$ in (7.2):

$$\begin{aligned} \left(\frac{d}{dt}\right)_{\text{fixed}} \left(\frac{d\vec{r}}{dt}\right)_{\text{rot}} &= \left(\frac{d^2\vec{r}}{dt^2}\right)_{\text{rot}} + \vec{\omega} \times \left(\frac{d\vec{r}}{dt}\right)_{\text{rot}} \\ \left(\frac{d}{dt}\right)_{\text{fixed}} (\vec{\omega} \times \vec{r}) &= \left[\frac{d}{dt}(\vec{\omega} \times \vec{r})\right]_{\text{rot}} + \vec{\omega} \times (\vec{\omega} \times \vec{r}) \end{aligned} \tag{7.6}$$

Substituting (7.6) in (7.5) we obtain

$$\begin{aligned} \left(\frac{d\vec{u}}{dt}\right)_{\text{fixed}} \equiv \left(\frac{d^2\vec{r}'}{dt^2}\right)_{\text{fixed}} = \frac{d^2\vec{a}}{dt^2} &+ \vec{\omega} \times (\vec{\omega} \times \vec{r}) + \left(\frac{d\vec{\omega}}{dt}\right)_{\text{rot}} \times \vec{r} \\ &+ 2\vec{\omega} \times \left(\frac{d\vec{r}}{dt}\right)_{\text{rot}} + \left(\frac{d^2\vec{r}}{dt^2}\right)_{\text{rot}} \end{aligned} \tag{7.7}$$

Note that the first three terms in the right-hand side of (7.7) survive even when $P$ is stationary relative to the rotating frame

$$S: \left(\frac{d\vec{r}}{dt}\right)_{\text{rot}} = \left(\frac{d^2\vec{r}}{dt^2}\right)_{\text{rot}} = 0 \tag{7.8}$$

Ignoring the $\frac{d^2\vec{a}}{dt^2}$ term by assuming that the distance $O\vec{O}'$ is not changing with time and observing that in an inertial frame Newton's second law implies that

$$m\left(\frac{d\vec{u}}{dt}\right)_{\text{fixed}} = \vec{F} \tag{7.9}$$

with $\vec{F}$ representing the vector sum of the forces acting on the particle, we have from (7.7)

$$m\left(\frac{d\vec{v}}{dt}\right) = \vec{F} - 2m(\vec{\omega} \times \vec{v}) - m\vec{\omega} \times (\vec{\omega} \times \vec{r}) - m\dot{\vec{\omega}} \times \vec{r} \tag{7.10}$$

where we have set

$$\vec{v} = \left(\frac{d\vec{r}}{dt}\right)_{\text{rot}}, \quad \frac{d\vec{v}}{dt} = \left(\frac{d^2\vec{r}}{dt^2}\right)_{\text{rot}} \quad \text{and} \quad \dot{\vec{\omega}} = \left(\frac{d\vec{\omega}}{dt}\right)_{\text{rot}} \tag{7.11}$$

and dropped the suffixes "fixed" and "rot" without causing any confusion. It should be clear that $\vec{v}$ is the velocity relative to the rotating frame.

Equation (7.10) is similar in form to Newton's equation. However, apart from $\vec{F}$, its right-hand side, is also influenced by fictitious forces, namely, the Coriolis force $-2m(\vec{v} \times \vec{\omega})$, the centrifugal force $-m\vec{\omega} \times (\vec{r} \times \vec{\omega})$ and the transverse or azimuthal force due to the non-uniform rotation $-m\dot{\vec{\omega}} \times \vec{r}$. The

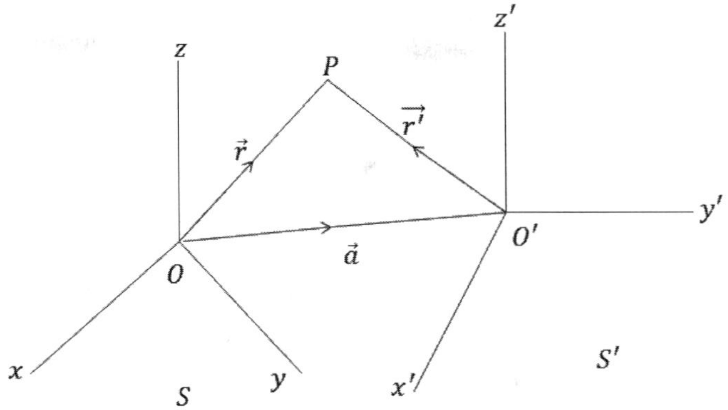


**FIGURE 7.3**: Typical three-dimensional motion with rotation about the $z-$axis.

latter can be neglected for a uniform rotation. Note that a fictitious force is due to an accelerated frame of reference. The form of the Coriolis term implies that it is always perpendicular to the direction of velocity and so it can never change the speed of a particle (except, of course, for its direction). As a result, the Coriolis force does not contribute to the energy equation. The Coriolis force is also referred to as a deflecting force.

Writing $\vec{F} = -\frac{\partial V}{\partial \vec{r}}$, where $V(\vec{r})$ is the potential, we can easily verify that the Lagrangian for (7.10) is

$$L = \frac{1}{2}m[v^2 + 2\vec{v}\cdot(\vec{\omega}\times\vec{r}) + (\vec{\omega}\times\vec{r})^2] - U \tag{7.12}$$

in which the velocity-dependent potential U is given by

$$U = V(\vec{r}) - \vec{v}\cdot(\vec{\omega}\times\vec{r}) - \frac{1}{2}m(\vec{\omega}\times\vec{r})^2 \tag{7.13}$$

Consider a typical three-dimensional motion with rotation about the $z-$axis (see Figure 7.3) and assuming $\vec{\omega}$ constant $(= \omega)$:

$$\vec{\omega} = \dot{\phi}\vec{k} \tag{7.14}$$
$$= \omega(\cos\theta\hat{e}_r - \sin\theta\hat{e}_\theta) \tag{7.15}$$

where $\hat{e}_r, \hat{e}_\theta$ are the unit basis vectors in the $\vec{r}$ and $\vec{\theta}$ directions, respectively.

Noting that

$$\begin{aligned} \vec{v} &= \dot{r}\hat{e}_r + (r\dot{\theta})\hat{e}_\theta \\ \frac{d\vec{v}}{dt} &= (\ddot{r} - r\dot{\theta}^2)\hat{e}_r + \frac{1}{r}\frac{d}{dt}(r^2\dot{\theta})\hat{e}_\theta \\ \vec{\omega}\times\vec{r} &= \omega(\cos\theta\hat{e}_r - \sin\theta\hat{e}_\theta)\times r\hat{e}_r = r\sin\theta\omega\hat{e}_\phi \end{aligned} \tag{7.16}$$

where $\hat{e}_\phi$ is the basis vector in the $\phi-$direction, we can write

$$\vec{\omega} \times \vec{v} = \omega(\dot{r}\sin\theta + r\dot{\theta}\cos\theta)\hat{e}_\phi \tag{7.17}$$

$$\vec{\omega} \times (\vec{\omega} \times \vec{r}) = -r\sin\theta(\sin\theta\hat{e}_r + \cos\theta\hat{e}_\theta)\omega^2 \tag{7.18}$$

Further since $\vec{u} = \vec{v} + \vec{\omega} \times \vec{r}$, we have componentwise

$$u_r = \dot{r}, \quad u_\theta = r\dot{\theta}, \quad u_\phi = r\sin\theta\dot{\phi} \tag{7.19}$$

Substituting (7.16), (7.17), (7.18), (7.19) in (7.10) we arrive at the following expressions of the various components of acceleration in a non-inertial rotating frame:

$$\begin{aligned}
\text{radial component of acceleration} \quad f_r &= \ddot{r} - r\dot{\theta}^2 - r\sin^2\theta\omega^2 \\
\text{cross-radial component of acceleration} \quad f_\theta &= \frac{1}{r}\frac{d}{dt}(r^2\dot{\theta}) - r\omega^2\sin\theta\cos\theta \\
\text{azimuthal component of acceleration} \quad f_\phi &= \frac{\omega}{r\sin\theta}\frac{d}{dt}(r^2\sin^2\theta)
\end{aligned} \tag{7.20}$$

### 7.1.2 Some remarks on the Coriolis force

From (7.10) the Coriolis force term is $-2m\vec{\omega} \times \vec{v}$ where $\vec{v} \equiv \frac{d\vec{r}}{dt}$. The Coriolis force is due to the rotation of the Earth. Consider the case of a flat rotating disc. For a particle moving across a disc under no forces, an inertial observer (i.e., the one who is in a fixed frame) will see it moving across a straight line (Figure 7.4) according to Newton's first law. However, in view of the fact that the disc is rotating, an observer stationed on the disc will view the particle taking a curved track (see Figure 7.5) due to the Coriolis

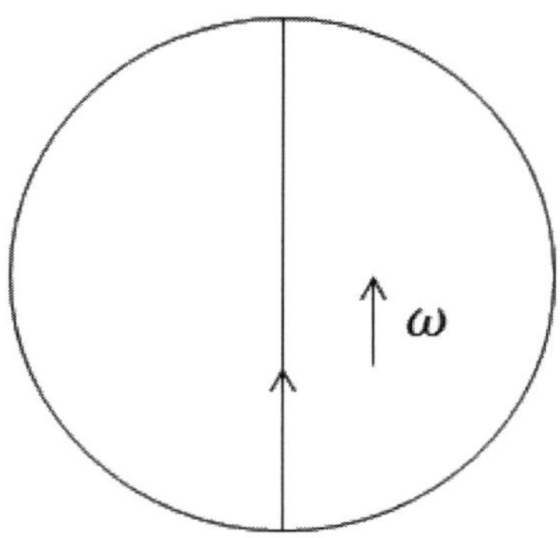


**FIGURE 7.4**: A particle moving across a disc under no forces.

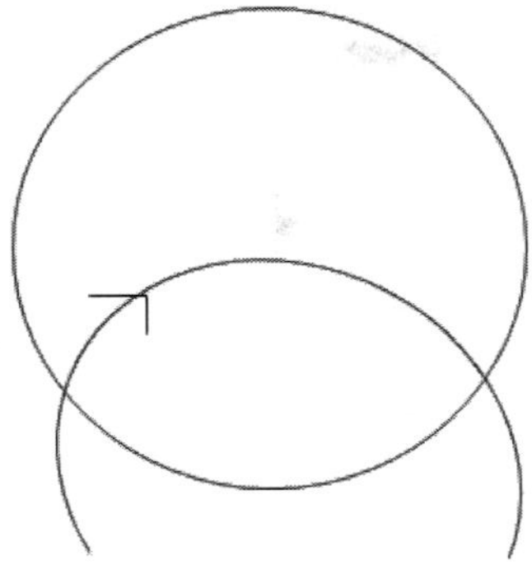

**FIGURE 7.5**: An observer stationed on the disc.

force operating in a direction perpendicular to the motion of the particle. Note that the Coriolis force causes the velocity vector to rotate clockwise in the Northern hemisphere and counterclockwise in the southern hemisphere. The Coriolis force causes hurricanes, tropical cyclones and such severe storms to rotate in different directions in the Northern and Southern hemispheres. The effect of the Coriolis force is minimum near the equator (in fact, it is zero there and hence its effect is more profound for the longitudinal motion, i.e., North-South wind, than the latitudinal counterpart, i.e., East-West wind) and maximum toward the poles. Thus a hurricane will never happen near the equator but gain in intensity as it approaches the poles.

### 7.1.3 Effective gravitational constant

Assuming $\vec{\omega}$ to be constant and neglecting the $\frac{d^2\vec{a}}{dt^2}$ term we can rewrite Equation (7.7) as

$$\left(\frac{d^2\vec{r'}}{dt^2}\right)_{\text{fixed}} = \left(\frac{d^2\vec{r}}{dt^2}\right)_{\text{rot}} + 2\vec{\omega} \times \left(\frac{d\vec{r}}{dt}\right)_{\text{rot}} + \vec{\omega} \times (\vec{\omega} \times \vec{r}) \tag{7.21}$$

For a particle moving under gravitation (i.e., the attractive force due to Newton's law of gravitation) and also subjected to a non-gravitational force $\vec{F}$, the equation of motion is

$$m\left(\frac{d^2\vec{r'}}{dt^2}\right)_{\text{fixed}} = m\vec{g} + \vec{F} \tag{7.22}$$

Thus from (7.21) and (7.22) we find

$$m\left(\frac{d^2\vec{r}}{dt^2}\right)_{\text{rot}} = m\vec{g} + \vec{F} - 2m\vec{\omega} \times \left(\frac{d\vec{r}}{dt}\right)_{\text{rot}} - m\vec{\omega} \times (\vec{\omega} \times \vec{r}) \tag{7.23}$$

In the laboratory when we measure the acceleration due to gravity what we determine is actually $\vec{g^*}$ which is the effective gravitational constant defined by the difference

$$\vec{g^*} = \vec{g} - \vec{\omega} \times (\omega \vec{\times} \vec{r}) \tag{7.24}$$

The right-hand side of (7.24) is a combination of gravitational and centrifugal forces. Thus a ball released near the Earth's surface will fall with acceleration $\vec{g^*}$. Because of the negative sign of the vector $-\vec{\omega} \times (\omega \vec{\times} \vec{r})$ which points radially outward, $\vec{g^*}$ in the Northern hemisphere will point to the south of the Earth's center.

The horizontal and vertical components of $\vec{g^*}$ are $g^*_{\text{hor}} = \omega^2 r \sin\theta \cos\theta$ and $g^*_{\text{ver}} = g - \omega^2 r \sin^2\theta$. At the pole $g^* = g$ while on the equator $g^* = g - \omega^2 r$.

Substituting (7.24) in (7.23) and writing $\vec{g}$ in place of $\vec{g^*}$ we have the formula

$$m\frac{d^2\vec{r}}{dt^2} = m\vec{g} + \vec{F} - 2m\vec{\omega} \times \frac{d\vec{r}}{dt} \tag{7.25}$$

where the suffix "rot" is dropped. The components of $\vec{\omega}$ being $(0, \omega\sin\theta, \omega\cos\theta)$, where $\theta$ is the angle between the direction of Earth's axis and $\vec{g^*}$, the Coriolis force is given by

$$-2m\vec{\omega} \times \frac{d\vec{r}}{dt} = 2m\omega(\dot{y}\cos\theta - \dot{z}\sin\theta, -\dot{x}\cos\theta, \dot{x}\sin\theta) \tag{7.26}$$

where $\frac{d\vec{r}}{dt} \equiv (\dot{x}, \dot{y}, \dot{z})$.

### 7.1.4 Foucault's pendulum

A useful device for observing the effects of the Coriolis force is the so-called Foucault's pendulum. The latter is a perfectly symmetric setup with the bob which is hanging from a string of length $l$ can swing freely in any vertical plane. Because of its symmetric nature, the periods of oscillation of Foucault's pendulum are equal in all directions. Neglecting the vertical component of the Coriolis force, which is very small compared to $g$, the equations of motion in the $x$ and $y$ directions are

$$\begin{aligned} \ddot{x} &= -\frac{g}{l}x + 2\omega\dot{y}\cos\theta \\ \ddot{y} &= -\frac{g}{l}y - 2\omega\dot{x}\cos\theta \end{aligned} \tag{7.27}$$

where we have used (7.25) and (7.26). One can see that the right-hand side of Equation (7.27) carries the contributions from the Coriolis acceleration in addition to the usual ones of the simple pendulum motion. In writing down (7.27) we also assumed that for small amplitude, the motion is nearly horizontal, i.e., $\dot{z} = 0$.

The underlying Lagrangian for the planar motion (7.27) is given by the form

$$L = \frac{1}{2}m(\dot{x}^2 + \dot{y}^2) + m\omega\cos\theta(x\dot{y} - y\dot{x}) - \frac{mg}{2l}(x^2 + y^2) \tag{7.28}$$

A straightforward way to solve (7.27) is to set $r = x + iy$ that results in the following complex linear equation:

$$\ddot{r} + 2ik\dot{r} + \omega_0^2 r = 0 \tag{7.29}$$

where $k = \omega\cos\theta \equiv \omega_\perp$ and $\omega_0^2 = \frac{g}{l}$.

Seeking a periodic solution $r = Ae^{-i\lambda t}$ where $A$ and $\lambda$ are constants leads to the constraint $\lambda^2 - 2k\lambda - \omega_0^2 = 0$. The two roots are given by the sum and difference

$$\lambda_\pm = k \pm \omega' \tag{7.30}$$

where $\omega'^2 = \omega_0^2 + \omega_\perp^2$. As a result we have for $r(t)$ the general expression

$$r(t) = c_1 e^{-i(k+\omega')t} + c_2 e^{-i(k-\omega')t} \tag{7.31}$$

where $c_1$ and $c_2$ are arbitrary complex constants which can be determined from the initial conditions of position and velocity.

An interesting particular case of (7.31) corresponds to the choice of the integration constants $c_1 = c_2 = \frac{1}{2}a$, in which the small transverse motion is omitted, and gives

$$\begin{aligned} r(t) &= ae^{-ikt}\cos\omega' t \\ \text{i.e.}\quad x(t) &= a\cos kt\cos\omega' t, \quad y(t) = -a\sin kt\cos\omega' t \end{aligned} \tag{7.32}$$

The square of the modulus of $r(t)$ turns out to be $|r(t)|^2 = a^2\cos^2\omega' t$. This speaks of a simple harmonic motion with amplitude $a$ and frequency $\omega' \simeq \omega_0 = \sqrt{\frac{g}{l}}$. The representations of (7.32) show that we have a superposition of slowly varying amplitudes given by the quantities $a\cos\omega_\perp t$ and $-a\sin\omega_\perp t$. During initial times (i.e., near $t = 0$) (or if $\omega_\perp$ being say close to the zero value near the equator where $\theta = \frac{\pi}{2}$), the oscillation of the pendulum tends to be simple harmonic entirely in the $x-$direction, i.e., in the north-south plane. However, with passage of time (additionally $\omega_\perp$ can acquire nonzero values), the amplitude of the $y-$coordinate grows reflecting an oscillation that has picked up an east-west component.

Finally from (7.32) we can determine the angle of inclination of the motion by evaluating the ratio $\tan\phi \equiv \frac{y(t)}{x(t)}$ which turns out to be $-\tan\omega_\perp t$ implying $\phi = -\omega_\perp t$. It means that the pendulum is rotating with an angular velocity $-\omega_\perp$ which is opposite to the direction of the rotation of the Earth. More specifically, in the Northern hemisphere where $\omega_\perp > 0$, we have a clockwise rotation of the plane of oscillation of the Foucault pendulum while in the Southern hemisphere it is just the opposite: the rotation of the plane of oscillation takes place in the counterclockwise direction.

## 7.2 Nonpotential force

We now consider a situation in which a physical system is acted upon by nonpotential forces denoted by $Q_j^{'}$:

$$Q_j^{'} = Q_j^{'}(q_k, \dot{q_k}, t), \quad j, k = 1, 2, ...n \tag{7.33}$$

in addition to the potential forces. It is clear that the nonpotential forces depend on generalized velocities for otherwise they will come in the category of potential forces. Indeed in the presence of (7.33) the generalized forces are to be expressed as

$$Q_j = -\frac{\partial V}{\partial q_j} + Q_j^{'} \tag{7.34}$$

where $V$ is the potential. As such Lagrange's equations of motion take the form

$$\frac{d}{dt}\left(\frac{\partial T}{\partial \dot{q_j}}\right) - \frac{\partial T}{\partial q_j} = -\frac{\partial V}{\partial q_j} + Q_j^{'}, \quad j = 1, 2, ...n \tag{7.35}$$

We look at the total energy $E = T + V$ where the kinetic energy $T$ is of the form $T = T(q_j, \dot{q_j}, t)$. The time rate of change of $T$ is

$$\begin{aligned}
\frac{dT}{dt} &= \sum_{j=1}^{n}\left(\frac{\partial T}{\partial q_j}\dot{q_j} + \frac{\partial T}{\partial \dot{q_j}}\ddot{q_j}\right) + \frac{\partial T}{\partial t} \\
&= \frac{d}{dt}\sum_{j=1}^{n}\frac{\partial T}{\partial \dot{q_j}}\dot{q_j} + \sum_{j=1}^{n}\left[\frac{\partial T}{\partial q_j} - \frac{d}{dt}\left(\frac{\partial T}{\partial \dot{q_j}}\right)\right]\dot{q_j} + \frac{\partial T}{\partial t} \\
&= \frac{d}{dt}\sum_{j=1}^{n}\frac{\partial T}{\partial \dot{q_j}}\dot{q_j} + \sum_{j=1}^{n}\left(\frac{\partial V}{\partial q_j} - Q_j^{'}\right)\dot{q_j} + \frac{\partial T}{\partial t}
\end{aligned} \tag{7.36}$$

where we have used (7.35)

As we learned in Chapter 3 the kinetic energy $T$ can be split up as

$$T = T_0 + T_1 + T_2 \tag{7.37}$$

where $T_2, T_1$ are quadratic and linear in generalized velocities, respectively, while $T_0$ is independent of it. Therefore, by Euler's theorem of homogeneous functions

$$\begin{aligned}
\sum_{j=1}^{n}\frac{\partial T_2}{\partial \dot{q_j}}\dot{q_j} &= 2T_2 \\
\sum_{j=1}^{n}\frac{\partial T_1}{\partial \dot{q_j}}\dot{q_j} &= T_1
\end{aligned} \tag{7.38}$$

As a result (7.36) takes the form

$$\begin{aligned}
\frac{dT}{dt} &= \frac{d}{dt}(2T_2+T_1)+\sum_{j=1}^{n}\left(-\frac{\partial V}{\partial q_j}-Q_j^{'}\right)\dot{q}_j+\frac{\partial T}{\partial t} \\
&= 2\frac{dT}{dt}-\frac{d}{dt}(T_1-2T_0)+\frac{\partial T}{\partial t}+\frac{dV}{dt}-\frac{\partial V}{\partial t}-\sum_{j=1}^{n}Q_j^{'}\dot{q}_j \quad (7.39)
\end{aligned}$$

where we have used $\frac{dV}{dt}=\sum_{j=1}^{n}\frac{\partial V}{\partial q_j}\dot{q}_j+\frac{\partial V}{\partial t}$ and also $\frac{\partial T_0}{\partial \dot{q}_j}=0$. The above equation implies that the time rate of change of the total energy is

$$\frac{dE}{dt}=\sum_{j=1}^{n}Q_j^{'}\dot{q}_j+\frac{d}{dt}(T_1+2T_0)-\frac{\partial}{\partial t}(T-V) \quad (7.40)$$

For a scleronomic system $T_1=T_0=0$ and $\frac{\partial T}{\partial t}=0$. If the potential energy too is not explicitly time-dependent then $\frac{dE}{dt}=\sum_{j=1}^{n}Q_j^{'}\dot{q}_j$. The latter is called the power of the nonpotential forces. Nonpotential forces are called gyroscopic if the power is zero: $\sum_{j=1}^{n}Q_j^{'}\dot{q}_j=0$ and dissipative if the power is negative $\sum_{j=1}^{n}Q_j^{'}\dot{q}_j<0$. In the former we have systems for which the nonpotential forces do not consume power while in the latter case the systems consume power. In dissipative systems, dissipative forces like friction are included, even through they sometimes do no work. The energy is generally lost through heat, sound, etc.

On the other hand, for a scleronomic system, the Coriolis force is a gyroscopic force. From the form (7.26) it is clear that since the Coriolis force for the $j^{th}$ particle is

$$\vec{F}_j^{cor}=-2m_j\left(\vec{\omega}\times\frac{d\vec{r_j}}{dt}\right)=-2m(\vec{\omega}\times\vec{v_j}) \quad (7.41)$$

we therefore have

$$\sum_{j=1}^{n}\vec{F}_j^{cor}.\vec{v_j}=0 \quad (7.42)$$

We conclude that the Coriolis force is a gyroscopic force.

---

## 7.3 Summary

This chapter was devoted to the investigation of motion in a noninertial coordinate system. After dealing with the general form of the equation of motion for rotating frames, the roles of the Coriolis force and effective gravitational constant were reviewed. We also discussed Foucault's pendulum which is a device for observing the effects of the Coriolis force and noted the typical implications that follow. Finally we dealt with the situation when a system is acted upon by nonpotential forces.

# Exercises

---

## 7.4 Examples

1. Find the deflection of a freely falling body from the vertical caused by Earth's rotation.

Solution: For a particle dropped from rest from a height $h$ above the ground the motion is described by the equations

$$x = 0, \quad y = 0, \quad z = h - \frac{1}{2}gt^2$$

where we have only considered gravity and neglected any effect of the Coriolis force. The time that the particle will take to hit the ground is $t = \sqrt{\frac{2h}{g}}$. The corresponding velocities are

$$\dot{x} = 0, \quad \dot{y} = 0, \quad \dot{z} = -g$$

To find the effect of Coriolis force on the equation of motion (7.25) we make use of these velocities in (7.26) to arrive at the following component equations:

$$m\ddot{x} = 2m\omega gt\sin\theta, \quad m\ddot{y} = 0, \quad m\ddot{z} = -gt$$

Using the above value of $t$ we find on integration of $m\ddot{x}$ the result

$$\begin{aligned} x &= \omega g\frac{1}{3}\sin\theta\left(\frac{2h}{g}\right)^{3/2} \\ \text{i.e.}\quad x &= \frac{1}{3}\omega\left(\frac{8h^3}{g}\right)^{1/2}\sin\theta \end{aligned}$$

where we used $x = 0, \quad \dot{x} = 0$ at $t = 0$.

We thus conclude that there will be an easterly deviation from the vertical by an amount $\frac{1}{3}\omega\left(\frac{8h^3}{g}\right)^{1/2}\sin\theta$.

2. A bead of mass $m$ slides freely along a smooth circular wire which is rotating with an angular velocity $\vec{\omega}$ about its fixed vertical diameter. Derive

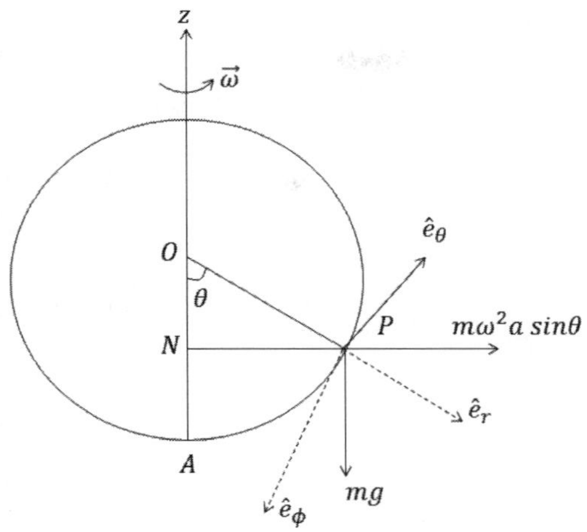


**FIGURE 7.6**: Motion of bead sliding along a smooth circular wire rotating with angular velocity $\omega$.

and discuss the energy conservation equation. Take $a$ to be the radius of the wire and set $\omega = \sqrt{\frac{ng}{a}}$, where $n$ is a parameter.

Solution: As shown in Figure 7.6 we have

$$\vec{\omega} = \omega \hat{e}_z = \omega(-\cos\theta \hat{e}_r + \sin\theta \hat{e}_\theta)$$

The centrifugal force has the magnitude $= m\omega^2 \left|\vec{NP}\right| = m\omega^2 a \sin\theta$ and acts horizontally outward along $\left|\vec{NP}\right|$. Obviously this force has a relevance in rotating frames only.

From the second relation of (7.20), the cross-radial equation of motion reads

$$m.\frac{1}{a}\frac{d}{dt}(a^2\dot{\theta}) = -mg\sin\theta + (m\omega^2 a\sin\theta)\cos\theta$$

$$\text{or,}\quad a\ddot{\theta} = -g\sin\theta + ng\sin\theta\cos\theta$$

where we have put $a\omega^2 = ng$. Integration gives

$$\frac{1}{2}ma^2\dot{\theta}^2 + \frac{mga}{2n}(1 - n\cos\theta)^2 = \text{constant}$$

This is essentially the energy conservation equation in a rotating frame with $\frac{1}{2}ma^2\dot{\theta}^2$ representing the kinetic energy and the potential energy is given by

$$V(n,\theta) = \frac{mga}{2n}(1 - n\cos\theta)^2$$

Differentiations yield

$$\begin{aligned} \frac{dV}{d\theta} &= mga\sin\theta(1 - n\cos\theta) \\ \frac{d^2V}{d\theta^2} &= mga[\cos\theta(1 - n\cos\theta) + n\sin^2\theta] \end{aligned}$$

Therefore $\frac{dV}{d\theta} = 0$ at $\theta = 0$ (namely, the lowest point $A$) and also at $\cos\theta = \frac{1}{n}$ for $n > 1$. Now a minimum of $V$ corresponds to the position of relative stable equilibrium and a maximum of $V$ corresponds to the position of relative unstable equilibrium.

**Case 1:** $\theta = 0$

Here

$$\begin{aligned} \frac{d^2V}{d\theta^2} &= mga(1 - n) > 0 \quad \text{if} \quad n < 1 \\ \Rightarrow \quad \theta &= 0 \quad \text{for} \quad n < 1 \quad \text{is a minimum for} \quad V \end{aligned}$$

Hence $\theta = 0$ is a position of relative stable equilibrium if $n < 1$. If however $n > 1$ then $\theta = 0$ is a maximum for $V$ and so $\theta = 0$ is a position of relative unstable equilibrium.

**Case 2:** $\cos\theta = \frac{1}{n} \quad (n > 1)$

We have

$$\frac{d^2V}{d\theta^2} = n\left(1 - \frac{1}{n^2}\right) > 0$$

Therefore $\cos\theta = \frac{1}{n}$ is a position of relative stable equilibrium. We conclude that new stable solutions are created at $\cos\theta = \frac{1}{n}$ as $n$ exceeds beyond the critical value of $n = 1$. Such a phenomenon is an example of bifurcation.

# Chapter 8

## Symmetries and conserved quantities



Symmetries can be characterized[1] in many ways depending on the problem context. They can be continuous (translations, rotations or a boost are some of the examples) which can be obtained by successive applications of infinitesimal transformations or discrete (like space reflection or time reversal for both of which Newtonian mechanics is invariant). They can be global (depending on parameters which are constant) or local (having functional dependence upon space and time). Symmetries can also be distinguished as those of geometrical or dynamical types. Examples of a geometrical symmetry are homogeneity and isotropy which arise due to the inner structure of spacetime. On the other hand, we can think of symmetries as being produced by the form invariance of the underlying equations. Dynamical symmetries are associated with such invariances.

Let us look at the following example of the gravitational equation of motion:

$$\ddot{\vec{r_j}} = -\sum_i \frac{Gm_i(\vec{r_j} - \vec{r_i})}{|\vec{r_j} - \vec{r_i}|^3}, \quad i = 1, 2, ..., n \tag{8.1}$$

where $m_i$ and $\vec{r_i}$ are the mass and position of the $i$-th particle while $G$ is Newton's constant. Consider the Galilean transformation

$$\vec{r_k} \to \vec{r'_k} = \vec{r_k} - \vec{V}t \tag{8.2}$$

where $\vec{V}$ is a constant velocity. We easily see that under such a transformation the form of the above equation remains unchanged:

$$\ddot{\vec{r'_j}} = -\sum_i \frac{Gm_i(\vec{r'_j} - \vec{r'_i})}{|\vec{r'_j} - \vec{r'_i}|^3}, \quad i = 1, 2, ..., n \tag{8.3}$$

[1]K. Brading and E. Castellani, Symmetry in classical physics, *Handbook of the Philosophy of Physics*, North-Holland, pp. 1331-1367, J. Butterfield and J. Earman, Eds.

This is a typical case of a dynamical symmetry. In the following we will first investigate the situation when a system exhibits continuous symmetry and proceed to show that it can then have an associated conserved quantity or charge. Discrete symmetries will be considered later.

---

## 8.1 Condition of invariance and Noether's theorem

In classical mechanics, Noether's theorem occupies a prominent position because according to this theorem if a symmetry is found to exist in a dynamical problem then there is a corresponding constant of motion. It provides a connection between global continuous symmetry and the resulting conservation law. The theorem for such an assertation was put forward by Emma Noether in the paper *Invariante Variationsprobleme* which came out in 1918.

To have an idea of how the proposition goes, we will be concerned with the invariance of a Lagrangian $L(q_i, \dot{q}_i, t), i = 1, 2, ..., n$ (where $q_i$ and $\dot{q}_i$ are, respectively, the generalized coordinates and velocities of a point particle in a system with $n$ degrees of freedom and $t$ is the time variable) under a reversible variable spacetime transformation of the type

$$q_i' = q_i'(q_1, q_2, ..., q_n; t), \quad t' = t'(t), \quad i = 1, 2, ..., n \tag{8.4}$$

Let us suppose that the primed coordinates and their velocities generate a Lagrangian $L'(q_i', \dot{q}_i', t'), i = 1, 2, ..., n$, where $\dot{q}_i' \equiv \frac{dq_i'}{dt'}$ with respect to the new time variable $t'$:

$$L(q_i, \dot{q}_i, t) \rightarrow L(q_i', \dot{q}_i', t'), \quad i = 1, 2, ..., n \tag{8.5}$$

We deal with those classes of suitable symmetry transformations that leave the form of the equations of motions invariant in the new primed system. Such transformations are referred to as the invariance transformations.

We already encountered in Chapter 3 the form invariance of equations of motion when, for instance, $L'$ differs from $L$ by a total time derivative term $\frac{d\Lambda}{dt}$, where $\Lambda$ is an arbitrary function of coordinates and time but independent of velocities. Expressed in terms of primed coordinates we noticed that corresponding to a relation

$$L'(q_i', \dot{q}_i', t') = L(q_i', \dot{q}_i', t') + \frac{d\Lambda(q_i', t')}{dt'}, \quad i = 1, 2, ..., n \tag{8.6}$$

the term $\frac{d\Lambda}{dt'}$ gives a vanishing contribution to Lagrange's equations of motion because of the following set of relations:

$$\frac{d}{dt}\left[\frac{\partial}{\partial \dot{q}'}\left\{\frac{d\Lambda(q',t)}{dt}\right\}\right] = \frac{\partial}{\partial q'}\left\{\frac{d\Lambda(q',t)}{dt}\right\}$$

or
$$\frac{d}{dt}\left[\frac{\partial}{\partial \dot{q}'}\left(\frac{\partial\Lambda}{\partial q'}\dot{q}' + \frac{\partial\Lambda}{\partial t}\right)\right] = \frac{\partial}{\partial q'}\frac{d\Lambda}{dt}$$

or
$$\frac{d}{dt}\left(\frac{\partial\Lambda}{\partial q'}\right) = \frac{\partial}{\partial q'}\frac{d\Lambda}{dt}$$

Let us look at the action integrals $S$ and $S'$ induced, respectively, by the Lagrangians $L$ and $L'$:

$$S = \int L(q_i, \dot{q}_i, \tau)d\tau, \quad S' = \int L'(q_i', \dot{q}_i', \tau')d\tau', \quad i = 1, 2, ..., n \tag{8.7}$$

Invariance of the actions implies $S = S'$ and so we write

$$\delta S \equiv S' - S = \int L'(q_i', \dot{q}_i', \tau')d\tau' - \int L(q_i, \dot{q}_i, \tau)d\tau = 0 \qquad i = 1, 2, ..., n \tag{8.8}$$

Using now the form of $L'$ as given by (8.6) we have

$$\int L(q_i', \dot{q}_i', \tau')d\tau' - \int L(q_i, \dot{q}_i, \tau)d\tau + \int \frac{d\Lambda(q_i', \tau')}{d\tau'}d\tau' = 0 \quad i = 1, 2, ..., n \tag{8.9}$$

Seeking invariance of the action under small infinitesimal perturbations where the coordinates $q_i$, the velocities $\dot{q}_i$ and $\tau$ transform according to

$$q_i' = q_i + \Delta q_i, \quad \dot{q}_i' = \dot{q}_i + \Delta\dot{q}_i, \quad \tau' = \tau + \Delta\tau \quad i = 1, 2, ..., n \tag{8.10}$$

(8.9) gives on substituting for the primed quantities

$$\int L(q_i + \Delta q_i, \dot{q}_i + \Delta\dot{q}_i, \tau + \Delta\tau)\left(1 + \frac{d\Delta\tau}{d\tau}\right)d\tau - \int L(q_i, \dot{q}_i, \tau)d\tau$$
$$+ \int \frac{d\Delta\Lambda(q_i + \Delta q_i, \tau + \Delta\tau)}{d\tau}d\tau = 0 \tag{8.11}$$

where $i = 1, 2, ..., n$ and in the last term of the left-hand side only the lowest-order infinitesimal is kept. We have also replaced $\Lambda$ by $\Delta\Lambda$ to indicate that the Lagrangian changes by a total derivative term of infinitesimal order.

Further, at the same level, Taylor expansion leads to

$$\int \left[\frac{\partial L}{\partial q_i}\Delta q_i + \frac{\partial L}{\partial \dot{q}_i}\Delta\dot{q}_i + \frac{\partial L}{\partial \tau}\Delta\tau + \frac{d\Delta\tau}{d\tau}L + \frac{d\Delta\Lambda(q_i, \tau)}{d\tau}\right]d\tau = 0 \tag{8.12}$$

The validity of (8.12) is ensured if the following holds:

$$\left(\frac{\partial L}{\partial q_i}\Delta q_i + \frac{\partial L}{\partial \dot{q}_i}\Delta\dot{q}_i + \frac{\partial L}{\partial \tau}\Delta\tau + \frac{d\Delta\tau}{d\tau}L\right)d\tau = -\frac{d\Delta\Lambda(q_i, \tau)}{d\tau}d\tau \tag{8.13}$$

Observe that the left-hand side is required to vanish if we insist upon the form-invariance of the Lagrangian. For instance, consider a pure space translation by a constant amount $k_i$. Then this implies $\sum_{i=1}^{N} k_i \frac{\partial L}{\partial q_i} = 0$ implying $\sum_{i=1}^{N} p_i = 0$ which stands for the vanishing of the total momentum acting on an $N$-particle system. On the other hand, for a constant time translation one finds $\frac{\partial L}{\partial \tau} = 0$ pointing to the time-independence of the Lagrangian. Thus for the specific cases of a pure space translation and a constant time displacement we have constancy of linear momentum and Hamiltonian, respectively.

The constancy of the total angular momentum following from the rotational invariance of the Lagrangian can be readily derived for the general setting of a physical system of $N$ particles described by the Lagrangian

$$L = \sum_{i=1}^{N} \frac{1}{2} m_i |\dot{\vec{r}}_i|^2 - \frac{1}{2} \sum_{i=1}^{N} V_{ij} \left( |\vec{r}_i - \vec{r}_j| \right) \tag{8.14}$$

Here $L$ represents a closed system of $N$ interacting particles under the influence of a potential $V_{ij}$ which operates between the $i$-th and $j$-th particle.

Consider an infinitesimal rotation by a quantity $\delta\theta$ about the $z$-axis (note that in the following we do not distinguish between $\delta$ and $\Delta$ variations as $t$ is not explicitly involved in $L$). Then $x$ and $y$ coordinates change according to

$$\begin{aligned} x' &= x\cos\theta - y\sin\theta \to x - y\delta\theta \\ y' &= y\cos\theta + x\sin\theta \to y + x\delta\theta \end{aligned} \tag{8.15}$$

Hence for an infinitesimal rotation about the $z$-axis we have

$$\sum_{i=1}^{N} \left[ \frac{\partial L}{\partial \dot{x}_i}(-y_i\delta\theta) + \frac{\partial L}{\partial \dot{y}_i}(x_i\delta\theta) \right] = \sum_{i=1}^{N} [m_i(-y_i\dot{x}_i + x_i\dot{y}_i)\delta\theta] = l_z\delta\theta \tag{8.16}$$

where $l_z$ stands for the $z$-th component of the angular momentum. Infinitesimal rotations about the $x$- and $y$-axis can be similarly written down. When we consider rotations in all the three directions we have $\vec{r'} - \vec{r} = \delta\vec{r} = \delta\vec{\theta} \times \vec{r}$ implying

$$\sum_{i=1}^{N} \frac{\partial L}{\partial \dot{\vec{r}}_i}.\delta\vec{r}_i = \sum_{i=1}^{N} m_i \dot{\vec{r}}_i.\delta\vec{\theta} \times \vec{r}_i = \vec{l}.\delta\vec{\theta} \tag{8.17}$$

where $\vec{l} = \sum_{i=1}^{N} \vec{r}_i \times m_i \dot{\vec{r}}_i$ is the total angular momentum of the system and represents a conserved Noether quantity.

More generally, we make use of the non-commutativity of $\frac{d}{d\tau}$ and $\Delta$ operators through a relation already written down in (6.62), namely,

$$\frac{d}{dt}(\Delta q) = \Delta\dot{q} + \dot{q}\frac{d}{dt}(\Delta t) \tag{8.18}$$

and adopt a similar strategy for derivation as we did to get (6.52) to arrive at the constancy condition

$$\frac{\partial L}{\partial \dot{q}_i}\Delta q_i + \left(L - \dot{q}_i \frac{\partial L}{\partial \dot{q}_i}\right)\Delta t + \Delta\Lambda = \text{constant} \tag{8.19}$$

The above is the central result of Noether's theorem which states that associated with an infinitesimal invariance transformation there is a constant of motion and hence an associated conservation law exists.

With $\frac{\partial L}{\partial \dot{q}_i} = p_i$, the generalized momentum, (8.19) can also be interpreted in terms of the Hamiltonian $H$ of the system for a conservative, scleronomic system having its kinetic energy a homogeneous function of the velocities. As we saw earlier, in such a case the quantity $\dot{q}_i \frac{\partial L}{\partial \dot{q}_i} - L$ is a constant and equals to $H$. We thus have the result

$$p_i \Delta q_i - H\Delta t + \Delta\Lambda = \text{constant} \tag{8.20}$$

A few points are in order regarding the applicability of Noether's theorem. First, it is relevant to systems enjoying a continuous class of symmetry (such as rotations as considered above) rather than a discrete one like parity (which seeks invariance under space reflection that reverses the sign of the position and velocity of the particle). This means that even if dissipative systems exhibit continuous symmetries there may not be any corresponding conservation law. Second, it produces conservation laws as we saw it ought to because of the underlying relation (8.19). Third, it insists upon the existence of a Lagrangian for (8.19) to be a valid constraint.

**Example 8.1:**

Show that angular momentum is conserved for a Lagrangian given by

$$L = (\dot{\vec{r}})^2 + \vec{r}.\dot{\vec{r}} + (\vec{r})^2$$

Under a rotation we have

$$\vec{r} \to \delta\vec{\theta} \times \vec{r}, \quad \dot{\vec{r}} \to \delta\vec{\theta} \times \dot{\vec{r}}$$

We see that the variation of the Lagrangian vanishes

$$\begin{aligned} \delta L &= \frac{\partial L}{\partial \vec{r}}.\delta\vec{r} + \frac{\partial L}{\partial \dot{\vec{r}}}.\delta\dot{\vec{r}} \\ &= \left(2\vec{r} + \dot{\vec{r}}\right).\left(\delta\vec{\theta} \times \vec{r}\right) + \left(\vec{r} + 2\dot{\vec{r}}\right).\left(\delta\vec{\theta} \times \dot{\vec{r}}\right) \\ &= 0 \end{aligned}$$

because of the vector identity $\vec{a}.(\vec{b} \times \vec{c}) = \vec{b}.(\vec{c} \times \vec{a})$.

**Example 8.2:**

Consider the free particle problem described by the Lagrangian

$$L = \frac{1}{2}m\dot{x}^2$$

How does $L$ transform under $x \to x + \lambda(t)$?

Since here

$$x' = x + \lambda(t), \quad \dot{x}' = \dot{x} + \dot{\lambda}$$

the Lagrangian in primed coordinate reads

$$L'(x', \dot{x}', t) = \frac{1}{2}m(\dot{x}' - \dot{\lambda})^2 = L(x', \dot{x}', t) - m\dot{\lambda}\dot{x}' + \frac{1}{2}m\dot{\lambda}^2$$

For an invariant transformation there exists some function $\Lambda(x', t)$ such that

$$L'(x', \dot{x}', t) = L(x', \dot{x}', t) + \frac{d\Lambda}{dt}$$

Writing

$$\frac{d\Lambda}{dt} = \frac{\partial \Lambda}{\partial x'}\dot{x}' + \frac{\partial \Lambda}{\partial t}$$

and comparing the above two forms of $L'(x', \dot{x}', t)$ we find

$$\frac{\partial \Lambda}{\partial x'} = -m\dot{\lambda}, \qquad \frac{\partial \Lambda}{\partial t} = \frac{1}{2}m\dot{\lambda}^2$$

Consistency of the relations yields $\ddot{\lambda} = 0$. It implies $\lambda(t)$ to be linear in time, i.e.,

$$\begin{aligned} \lambda(t) &= c_1 + c_2 t \\ \Rightarrow \quad \Lambda(x, t) &= -mc_2 x' + \frac{1}{2}mc_2^2 t \end{aligned}$$

where $c_1$ and $c_2$ are arbitrary constants. Thus $x'$ is of the form $x' = x + c_1 + c_2 t$ which is a combination of spatial displacement ($c_2 = 0$) and Galilean transformation ($c_2 \neq 0$). It also follows that a transformation to a uniformly accelerating frame does not constitute an invariant transformation.

## 8.2 Operator approach

In this section we address the formulation of dynamical symmetries in terms of operators[2] that act on the state space for the system. A classical dynamical system is specified by a set of dynamical variables

$$S = f(r_1, r_2, ..., r_n) \tag{8.21}$$

Note that the time $t$ is not considered explicitly.

The set $S$ evolves in time according to the equations of motion

$$\dot{r_k} = \phi_k(r_1, r_2, ..., r_n; t), \quad k = 0, 1, 2, ..., n \tag{8.22}$$

where $\phi_k$'s are, as yet, some unspecified functions of coordinates and time. As an example, if n=2, we can identify $r_1 = z$ as the particle position and $r_2 = v$ as its velocity:

$$\dot{z} = v, \quad m\dot{v} = F(z, v, t) \tag{8.23}$$

where $F(z, v, t)$ is some force function which depends on the coordinate, velocity and time in a very general setting. For an autonomous system we have the reduced form

$$\dot{r_k} = \phi_k(r_1, r_2, ..., r_n), \quad k = 0, 1, 2, ..., n \tag{8.24}$$

We now inquire into the time development of $S$ from instants $t = t_i$ to $t = t_f$. Introducing the corresponding evolution operator $U(t_f, t_i)$, such an evolution can be expressed as

$$U(t_f, t_i)[r_1(t_i), ..., r_n(t_i)] = [r_1(t_f), ..., r_n(t_f)] \tag{8.25}$$

Should the equations of motion not depend upon time $t$ explicitly, it is evident that $U(t_f, t_i)$ would depend only on the difference $t_f - t_i$ and we denote in such a case

$$U = U(\tau) = U(\tau, 0), \quad \tau = t_f - t_i \tag{8.26}$$

We are then led to an invariance under time translation:

$$U(t_f + \Delta, t_i + \Delta) = U(t_f, t_i) \tag{8.27}$$

for a constant time interval $\Delta$.

Some simple expressions of the evolution operator resulting from an appropriate choice of the force function are given by the following examples:

[2] A.D. Boozer, Dynamical symmetries in classical mechanics, *Eur. J. Phys.* 33(2012) 73.

(1). The case of the free particle for which the force function $F(z; v) = 0$. Here $U(\tau)$ is given by

$$U(\tau)(z, v) = (z + v\tau, v)$$

(2). The case of simple harmonic motion described by $F(z; v) = -m\omega^2 z$ for which $U(\tau)$ is given by

$$U(\tau)(z, v) = (z \cos \omega\tau + \frac{v}{\omega} \sin \omega\tau, v \cos \omega\tau - z\omega \sin \omega\tau)$$

(3). The case of damped particle under the influence of $F(z; v) = -m\gamma v$ where $\gamma$ is a damping constant. Here $U(\tau)$ is given by

$$U(\tau)(z, v) = z + \frac{v}{\gamma}[(1 - e^{-\gamma\tau}), ve^{-\gamma\tau}]$$

(4). The case of a uniformly accelerated Newtonian particle with an acceleration $a$ subjected to the force function $F(z; v) = ma$. Here $U(\tau)$ is given by

$$U(\tau)(z, v) = (z + v\tau + \frac{a\tau^2}{2}, v + a\tau)$$

### 8.2.1 Symmetry operator

We can define for every symmetry transformation a corresponding time-dependent coordinate transformation operator $\xi(t)$ and consider the evolution of the set of dynamical variables $(r_1, r_2, ..., r_n)$ according to some given equations of motion. The form invariance of the latter requires that

$$\dot{r'_k} = \phi_k(r'_1, r'_2, ..., r'_n; t), \quad k = 0, 1, 2, ..., n \tag{8.28}$$

Transforming to new dynamical variables $(r'_1, ..., r'_n)$ under $\xi(t)$ we can write

$$\xi(t)(r_1, r_2..., r_n) = (r'_1, r'_2, ..., r'_n) \tag{8.29}$$

Assuming an inverse of $\xi(t)$ to exist, it follows that

$$\xi^{-1}(t)(r'_1, ..., r'_n) = (r_1, ..., r_n) \tag{8.30}$$

For the new set of dynamical variables $(r'_1, r'_2, ..., r'_n)$ let us define a corresponding evolution operator $U'(t_f, t_i)$ such that

$$U'(t_f, t_i)[r'_1(t_i), ..., r'_n(t_i)] = [r'_1(t_f), ..., r'_n(t_f)] \tag{8.31}$$

We therefore deduce

$$\begin{aligned} U'(t_f, t_i)\xi(t_i)[r_1(t_i), ..., r_n(t_i)] &= \xi(t_f)[r_1(t_f), ..., r_n(t_f)] \\ &= \xi(t_f)U(t_f, t_i)[r_1(t_i), ..., r_n(t_i)] \end{aligned} \tag{8.32}$$

where for the second equality we have used (8.25). Consistency requires

$$U'(t_f, t_i)\xi(t_i) = \xi(t_f)U(t_f, t_i) \tag{8.33}$$

yielding the following connection between $U'(t_f, t_i)$ and $U(t_f, t_i)$:

$$U'(t_f, t_i) = \xi(t_f)U(t_f, t_i)\xi^{-1}(t_i) \tag{8.34}$$

Form invariance of the equations of motion requires that the evolution operator for the old variables and the transformed ones be alike: $U(t_f, t_i) = U'(t_f, t_i)$ resulting in the invariance result

$$U(t_f, t_i) = \xi(t_f)U(t_f, t_i)\xi^{-1}(t_i) \tag{8.35}$$

For the time independent case when $\xi(t) = \xi$, we get the simple commutative condition

$$U(t_f, t_i)\xi = \xi U(t_f, t_i) \tag{8.36}$$

The operator method is a powerful approach for providing connections to discrete symmetries. Let us try for the operations of parity and time-reversal.

### 8.2.2 Parity transformation

The parity operator P is associated with reflection in space. It reverses the sign of the position and velocity of the particle:

$$P : z \to -z, \quad v \to -v \tag{8.37}$$

It is a discrete symmetry and for the invariance of a system under such an operation the commutative condition holds as given below:

$$U(t_f, t_i)P = PU(t_f, t_i) \tag{8.38}$$

A free particle is invariant under parity transformation as can be directly verified:

$$PU(\tau)[z, v] = P[z + v\tau, v] = [-z - v\tau, -v] \tag{8.39}$$

$$U(\tau)P[z, v] = P[-z, -v] = [-z - v\tau, -v] \tag{8.40}$$

However in the case of a Newtonian particle which is uniformly accelerated, the symmetry under parity transformation is not preserved:

$$PU(\tau)[z,v] = P[(z + v\tau + \frac{a\tau^2}{2}, v + a\tau)] = [-z - v\tau - \frac{a\tau^2}{2}, -v - a\tau] \quad (8.41)$$

but

$$U(\tau)P[z,v] = U[-z,-v] = [(-z - v\tau + \frac{a\tau^2}{2}, -v + a\tau)] \quad (8.42)$$

### 8.2.3 Time-reversal symmetry

The time-reversal operator reverses the sign of the time variable $t$ according to

$$T : t \to -t \quad (8.43)$$

Like the parity, the time-reversal transformation is discontinuous too and so not amenable to the treatment of Section 8.1. Note that the discontinuous character means that the underlying transformation cannot be developed from a series of infinitesimal transformations as is true for the continuous case.

If the system undergoes a time-reversal change, the transformed equations of motion will be given by a similar set as in (8.24):

$$\dot{r'_k} = -\phi_k(r'_1, r'_2, ..., r'_n), \quad k = 0, 1, 2, ..., n \quad (8.44)$$

where because of no explicit time dependence $\phi_k$'s remain unchanged.

At this point if we integrate (8.24) between an initial time $t_i$ and a final time $t_f$, then an operator can be defined depending on $\tau$, which is the difference $t_f - t_i$, evolving the dynamical variables $r_1, r_2, ..., r_n$. On the other hand, if we integrate (8.44), we can think of another operator which evolves the variables $r'_1, r'_2, ..., r'_n$. We therefore have the connection $U'(\tau) = U(-\tau)$.

If the link between the two sets $(r'_1, r'_2, ..., r'_n)$ and $(r_1, r_2, ..., r_n)$ is provided by the application of an operator $T$ on the latter, i.e.,

$$T(r_1, r_2, ..., r_n) = (r'_1, r'_2, ..., r'_n) \quad (8.45)$$

then

$$U'(\tau) = TU(\tau)T^{-1} \quad (8.46)$$

We therefore arrive at the result

$$TU(\tau) = U(-\tau)T \quad (8.47)$$

Time reversal implies that the role of time is reversed, in other words, there is a reversal of motion. In Newtonian mechanics in which the second law is the same with regard to the explicit replacement $t \to -t$, time reversal means that the particle retraces its path and ultimately arrives at the point it started from with a reversed momentum. Note that time-reversal invariance is not obeyed in our previously considered example of a damped particle.

---

## 8.3 Virial theorem

Virial theorem gives a relation for the average over time of the total kinetic energy. The result can be derived as follows.

Consider an $N$-particle system as rest for which we define a quantity $G$

$$G = \sum_{i=1}^{N} \vec{p}_i . \vec{r}_i \tag{8.48}$$

where $\vec{r}_i$ denotes the $i^{th}$ position of a particle with respect to a fixed origin. Clearly $G$ is restricted to be finite since a material particle has neither an infinite momentum nor can it be infinitely away from the origin.

Translation of the origin by a finite amount $\vec{r}_0$ leaves $G$ invariant:

$$\begin{aligned} G' &= \sum_{i=1}^{N} \vec{p}_i . \left(\vec{r}_i - \vec{r}_0\right) \\ &= G - \vec{r}_0 \sum_{i=1}^{N} \vec{p}_i \\ &= G \end{aligned}$$

from the conservation of linear momentum.

The time rate of change of $G$ is given by

$$\begin{aligned} \dot{G} &= \sum_{i=1}^{N} \left( \dot{\vec{p}}_i . \vec{r}_i + \vec{p}_i \dot{\vec{r}}_i \right) \\ &= 2T + \sum_{i=1}^{N} \vec{F}_i . \vec{r}_i \\ &= 2T + W \end{aligned}$$

where $T$ is the kinetic energy and $W$ is called the virial of the system. Like $G$, $W = \sum_{i=1}^{N} \vec{F}_i . \vec{r}_i$ too is independent of the choice of origin for the system at rest.

Let us go for the time average of $2T+W$. We define

$$\begin{aligned}\Delta = <2T+W> &= \lim_{\tau\to\infty}\frac{1}{\tau}\int_0^\tau \dot{G}dt \\ &= \lim_{\tau\to\infty}\frac{G(\tau)-G(0)}{\tau} \\ &= 0\end{aligned}$$

since $G(\tau)-G(0)$ is always finite. Hence

$$\bar{T} = -\frac{1}{2}\bar{W} \tag{8.49}$$

This is the virial theorem.

Consider a two-particle system for which

$$W = \vec{f}_{12}.\vec{r}_2 + \vec{f}_{21}.\vec{r}_1 = \vec{f}_{12}.\left(\vec{r}_2 - \vec{r}_1\right) \tag{8.50}$$

where $\vec{f}_{ij}, i,j = 1,2$ are the inter-particle forces and we have used Newton's third law. If $V_{12} = -\frac{\gamma}{|\vec{r}_2-\vec{r}_1|}$ is the inter-particle potential ($\gamma$ a constant) then $\vec{f}_{12}$ is

$$\vec{f}_{12} = -\frac{\gamma(\vec{r}_2-\vec{r}_1)}{|\vec{r}_2-\vec{r}_1|^3} \tag{8.51}$$

It shows

$$\vec{f}_{12}.(\vec{r}_2-\vec{r}_1) = -\frac{\gamma}{|\vec{r}_2-\vec{r}_1|} = V_{12} \tag{8.52}$$

As such $W$ represents the total potential energy $V(=V_{12})$ of the system and implies

$$\bar{T} = -\frac{1}{2}\bar{W} = -\frac{1}{2}\bar{V} \tag{8.53}$$

From the conservation of total energy we can write the above result as

$$E = -\bar{T} = \frac{1}{2}\bar{V} \tag{8.54}$$

Virial theorem is of great importance in many branches of physics especially in statistical mechanics. It does not depend on the notion of the temperature and holds even for systems that are not in thermal equilibrium.

---

## 8.4 Summary

The purpose of this chapter was to investigate the role of symmetry, essentially the dynamical ones, in classical mechanics. We gave a simple derivation of the condition of invariance from which we extracted the essence of Noether's theorem which holds for a continuous class of symmetries. After working out a number of examples we discussed the problem of dynamical symmetry following an operator approach in which the notion of an evolution operator was employed. The operator method allows one to address discrete symmetries such as the parity and time-reversal operations. We also considered briefly the virial theorem.

## Exercises

1. Consider a ray of light traveling from the point $P_0$ to the point $P_1$ through the point of reflection at the point $(x, 0)$ on a mirror M. Calculate the time taken and hence derive the law of reflection.

2. Consider two mediums $S_0$ and $S_1$ characterized, respectively, by the index of refraction $n_0$ and index of refraction $n_1$. Calculate the time taken for the light traveling from the point $P_0$ to the point $P_1$ through the point $(x, 0)$. Hence, derive Snell's law of refraction.

3. Show that the shortest distance between two points in a plane is a straight line.

4. Show that the shortest distance between two points on the surface of a sphere is a great circle.

5. Find the equation of the curve which makes the surface area of revolution generated by rotating the curve $y = y(x)$ around the x-axis.

# Chapter 9

## Hamilton–Jacobi equation and action-angle variables



The analytical structure of classical dynamics describing the motion of a particle or a system of particles rests on two important but almost equivalent principles, namely, those of Lagrangian and Hamiltonian mechanics. Let us recall that for, say, a time-independent mechanical system of particles with $n$ degrees of freedom, the Lagrangian is a function of the generalized coordinates $q_i, i = 1, 2, ..., n$ and generalized velocities $\dot{q}_i, i = 1, 2, ..., n$ and appears as the difference of kinetic and potential energies of the particles. To get a knowledge of the dynamics of the motion requires solving $2n$ second-order differential equations subject to a set of appropriate initial conditions. Although the formulation of Lagrangian mechanics does not yield any additional information as compared to what is generally known from Newtonian mechanics, the main advantage with it is that one does not have to worry with the presence of constraint forces since the latter are eliminated by invoking D'Alembert's principle. If the system is not conservative, dissipative or driven forces are taken into account by the non-potential terms. One important aspect which makes a Lagrangian system very useful is the presence of conserved quantities whenever one or more generalized coordinates are absent from the Lagrangian. On the other hand, the Hamiltonian, which can be derived from the Lagrangian by effecting a Legendre transformation, is regarded as a function of the generalized coordinates $q_i, i = 1, 2, ..., n$ and generalized momenta $p_i, i = 1, 2, ..., n$, and appears, in a conservative system, as the sum of kinetic and potential energies. The underlying equations of motion are given by a set of $n$ first order partial differential equations each for the time derivative of the generalized coordinates and momenta. The Hamiltonian procedure is carried out in a phase space which is a $2n$-dimensional space made up of positions and the accompanying canonical momenta and the Hamiltonian equations give the clue as to how a system evolves in phase space. Both Lagrangian and Hamiltonian formalisms address the laws of mechanics without any preference to the selection of any particular coordinate system.

The Hamiltonian procedure, however, does not always offer any simplification to the concerned dynamical problem. But the almost symmetrical appearence of the coordinates and momenta in the guiding Hamilton's equations facilitates development of formal theories such as the canonical transformations, Hamilton–Jacobi equation and action-angle variables. Our first purpose in this chapter will be to understand how canonical transformations are constructed that offer considerable simplifications in writing down the equations of motion. The task is to transform the Hamiltonian to a new form through the introduction of a new set of coordinates and momenta such that with respect to the new set of variables the transformed Hamiltonian also leads to the form of canonical equations. The utility of canonical transformation lies in the fact that it is often possible to adopt new sets of conjugate variables through which the basic equations get much simplified thus generating solutions that may otherwise be very complicated to determine. The special case when all the coordinates are cyclic yields a single partial differential equation of first order called the Hamilton–Jacobi equation. In this chapter we also deal briefly with the elements of a certain class of canonical variables called the action-angle variables defined over a bounded phase space that emerge from a complete separability of the Hamilton–Jacobi equation. Some interesting properties of such variables are explored.

## 9.1 Canonical transformation

For simplicity we will be frequently restricting to a system of one degree of freedom described by a single pair of canonical variables $(q, p)$. The phase space is then of dimension two. The Hamiltonian equations are given by

$$\dot{q} = \frac{\partial H}{\partial p}, \quad \dot{p} = -\frac{\partial H}{\partial q} \tag{9.1}$$

We ask the question whether a transformation from $(q, p)$ to a new set of variables $(Q, P)$ is feasible in a way that the above form of the Hamiltonian equations is preserved

$$\dot{Q} = \frac{\partial K}{\partial P}, \quad \dot{P} = -\frac{\partial K}{\partial Q} \tag{9.2}$$

We emphasize that our intention in looking for new variables $(Q, P)$ is to inquire if these could also serve in a new but simplified means of describing the system but at the same time not disturb the essential physics content. In (9.2) $K$ is a transformed version of $H$

$$H(q(Q,P), p(Q,P)) \to K(Q,P) \equiv K \tag{9.3}$$

defined in terms of the variable $Q$ and $P$. It is obvious that the equations (9.2) may not always hold except for some special situations. Such restricted transformations for which (9.2) are valid are called canonical transformations implying that the new variables $(Q, P)$ too form a canonical set.

Noting that the Poisson bracket of any two functions $F$ and $G$, defined in terms of the old canonical variables $q$ and $p$, is given by

$$\{F, G\}_{(q,p)} = \frac{\partial F}{\partial q}\frac{\partial G}{\partial p} - \frac{\partial F}{\partial p}\frac{\partial G}{\partial q}$$

Hamiltonian canonical equations for $Q = Q(q, p), P = P(q, p)$ read

$$\begin{aligned} \dot{Q} &= \{Q, H\}_{(q,p)} = \frac{\partial Q}{\partial q}\frac{\partial H}{\partial p} - \frac{\partial Q}{\partial p}\frac{\partial H}{\partial q} \\ \dot{P} &= \{P, H\}_{(q,p)} = \frac{\partial P}{\partial q}\frac{\partial H}{\partial p} - \frac{\partial P}{\partial p}\frac{\partial H}{\partial q} \end{aligned} \tag{9.4}$$

Interpreting $H(q, p)$ as $K(Q, P)$ in terms of the new variables $(Q, P)$ and employing the chain rule of partial derivatives gives

$$\begin{aligned} \frac{\partial H}{\partial q} &= \frac{\partial K}{\partial Q}\frac{\partial Q}{\partial q} + \frac{\partial K}{\partial P}\frac{\partial P}{\partial q} \\ \frac{\partial H}{\partial p} &= \frac{\partial K}{\partial Q}\frac{\partial Q}{\partial p} + \frac{\partial K}{\partial P}\frac{\partial P}{\partial p} \end{aligned} \tag{9.5}$$

This results in the following expressions for $\dot{Q}$ and $\dot{P}$:

$$\begin{aligned}
\dot{Q} &= \frac{\partial Q}{\partial q}\left(\frac{\partial K}{\partial Q}\frac{\partial Q}{\partial p} + \frac{\partial K}{\partial P}\frac{\partial P}{\partial p}\right) - \frac{\partial Q}{\partial p}\left(\frac{\partial K}{\partial Q}\frac{\partial Q}{\partial q} + \frac{\partial K}{\partial P}\frac{\partial P}{\partial q}\right) \\
&= \frac{\partial K}{\partial P}\left(\frac{\partial Q}{\partial q}\frac{\partial P}{\partial p} - \frac{\partial Q}{\partial p}\frac{\partial P}{\partial q}\right) \\
&= \frac{\partial K}{\partial P}\{Q, P\}_{(q,p)} \\
\dot{P} &= -\frac{\partial K}{\partial Q}\{Q, P\}_{(q,p)}
\end{aligned} \tag{9.6}$$

These equations coincide with (9.2) provided we set $\{Q, P\}_{(q,p)}$ equal to unity:

$$\{Q, P\}_{(q,p)} = 1 \tag{9.7}$$

The Hamiltonian form of canonical equations are then valid for the new pair of variables $(Q, P)$.

For a system of $n$ degrees of freedom the condition for the canonical transformation is given by the Poisson bracket conditions

$$\{Q_i, Q_j\}_{(q,p)} = 0, \quad \{P_i, P_j\}_{(q,p)} = 0, \quad \{Q_i, P_j\}_{(q,p)} = \delta_{ij} \tag{9.8}$$

Invariance of Poisson bracket relations is a fundamental feature of canonical transformations. In fact the above properties of the Poisson brackets serve as the necessary and sufficient conditions for a transformation to be canonical. Note that out of $2n$ variables, $(q_i, p_i)$ and $(Q_i, P_i)$, $i = 1, 2, ..., n$, only $2n$ of these are independent.

It needs to be pointed out that the Poisson bracket $\{Q, P\}_{(q,p)}$ is the same as the Jacobian determinant:

$$\begin{aligned}
\{Q, P\}_{(q,p)} &= \frac{\partial Q}{\partial q}\frac{\partial P}{\partial p} - \frac{\partial Q}{\partial p}\frac{\partial P}{\partial q} \\
&= \frac{\partial(Q, P)}{\partial(q, p)}
\end{aligned} \tag{9.9}$$

Conversely we also have

$$\{q, p\}_{(Q,P)} = \frac{\partial(q, p)}{\partial(Q, P)} = \left[\frac{\partial(Q, P)}{\partial(q, p)}\right]^{-1} = \left[\{Q, P\}_{(q,p)}\right]^{-1} \tag{9.10}$$

Consider a region $D$ in the phase space plane $(q, p)$ that is bounded by a closed curve $A$. Using (9.9) we can express

$$\int_D dQdP = \int_D \frac{\partial(Q, P)}{\partial(q, p)} dqdp = \int_D [Q, P]_{q,p} dqdp = \int_D dqdp \tag{9.11}$$

where we have exploited (9.7). In terms of closed line integrals we then have by Stokes's theorem

$$\oint_A pdq = \oint_A PdQ \tag{9.12}$$

In general, for a system with $n$ generalized coordinates $q_i$ together with their associated momenta $p_i$, the volume element transforms as

$$\prod_{i=1}^{n} dQ_i dP_i = \Delta \prod_{i=1}^{n} dq_i dp_i = \prod_{i=1}^{n} dq_i dp_i \tag{9.13}$$

where $\Delta$ stands for the determinant of the Jacobian matrix

$$\Delta = \det[\frac{\partial(Q_1, Q_2, ..., Q_n; P_1, P_2, ..., P_n)}{\partial(q_1, q_2, ..., q_n; p_1, p_2, ..., p_n)}] \tag{9.14}$$

and we employed Liouville's theorem (see Chapter 4, Section 4.7). Thus phase space volume is preserved by canonical transformations.

**Example 9.1:**

Consider the transformation $(q, p) \to (Q, P)$ given by

$$q = \sqrt{\frac{2P}{m\omega}} \sin Q, \quad p = \sqrt{2m\omega P} \cos Q$$

Inverting

$$Q = \tan^{-1}\left(m\omega\frac{q}{p}\right), \quad P = \frac{1}{2\omega}\left(\frac{p^2}{m} + m\omega^2 q^2\right)$$

$$\Rightarrow \quad \frac{\partial Q}{\partial q} = \frac{m\omega p}{p^2 + m^2\omega^2 q^2}, \quad \frac{\partial Q}{\partial p} = -\frac{m\omega q}{p^2 + m^2\omega^2 q^2}$$

As a result

$$\begin{aligned}\{Q, P\}_{(q,p)} &= \frac{\partial Q}{\partial q}\frac{\partial P}{\partial p} - \frac{\partial Q}{\partial p}\frac{\partial P}{\partial q} \\ &= \frac{1}{p^2 + m^2\omega^2 q^2}\left(m\omega p\frac{\partial P}{\partial p} + m\omega q\frac{\partial P}{\partial q}\right)\end{aligned}$$

Since $\frac{\partial P}{\partial p} = \frac{p}{m\omega}$, $\frac{\partial P}{\partial q} = m\omega q$ it follows that $\{Q, P\}_{(q,p)} = 1$. Hence the transformation is canonical.

**Example 9.2:**

Let us focus on the specific case of the harmonic oscillator described by the Hamiltonian $H = \frac{1}{2m}p^2 + \frac{1}{2}m\omega^2 q^2$. It is clear from the form of $P$ in the previous example that if employed the Hamiltonian takes a very simple form $H \to K = P\omega$, where $K$ is the transformed Hamiltonian. The accompanying Hamilton's equations for the new variables $Q$ and $P$ are easily read off

$$\dot{Q} = \frac{\partial K}{\partial P} = \omega$$
$$\dot{P} = -\frac{\partial K}{\partial Q} = 0$$

Solving for $Q$ and $P$ we find $Q = \omega t + t_0, \quad P = b$ where $t_0$ and $b$ are constants of integration. Switching to the original variables we get for $q$ and $p$

$$q = \sqrt{\frac{2b}{m\omega}} \sin(\omega t + t_0)$$
$$p = \sqrt{2m\omega b} \cos(\omega t + t_0)$$

which conform to their standard forms.

This example serves to illustrate the advantage of employing canonical variables in a suitable manner as a result of which the basic equations get simplified and obtaining the solutions becomes an easy task.

---

## 9.2 Symplectic property

Let us point out that the canonical transformations obey the symplectic property. Indeed we observe that corresponding to the matrix of derivatives of the canonical variables $(Q, P)$ namely

$$J = \begin{pmatrix} \partial Q/\partial q & \partial Q/\partial p \\ \partial P/\partial q & \partial P/\partial p \end{pmatrix} \tag{9.15}$$

and in terms of an off-diagonal antisymmetric matrix $A$ defined by

$$A = \begin{pmatrix} 0 & 1 \\ -1 & 0 \end{pmatrix} \tag{9.16}$$

the relation $A = JAJ^T$, $J^T$ being the transpose of $J$, is always valid. Indeed a simple multiplication gives

$$JAJ^T = \begin{pmatrix} \{Q, Q\}_{(q,p)} & \{Q, P\}\{_{(q,p)} \\ -\{Q, P\}_{(q,p)} & \{P, P\}_{(q,p)} \end{pmatrix} \tag{9.17}$$

where the entries stand for the Poisson brackets. Using their properties as furnished in (9.8) our assertion is seen to hold. We refer to $J$ as a symplectic matrix and the canonical transformations induced by $(Q, P)$ are said to be symplectic.

In a general sense, Hamilton's equations for a $2n$-dimension phase space, defined by the generalized coordinates and generalized momenta which we represent by a column vector $\boldsymbol{X} \equiv (q_1, q_2, ..., q_n; p_1, p_2, ..., p_n)^T \equiv (q, p)^T$, can be written as

$$\dot{\boldsymbol{X}} = A\frac{\partial H}{\partial X_i} = A\nabla_X H(\boldsymbol{X}) \Rightarrow \quad \frac{d}{dt}\begin{pmatrix} q \\ p \end{pmatrix} = A\begin{pmatrix} \nabla_q H(q,p,t) \\ \nabla_p H(q,p,t) \end{pmatrix} \tag{9.18}$$

where $A$ is a $2n \times 2n$ symplectic matrix given by

$$A \;=\; \begin{pmatrix} 0 & +I_n \\ -I_n & 0 \end{pmatrix} \tag{9.19}$$

with $I_n$ representing the $(n \times n)$ block identity matrix. The properties of $A$ are given by $(A)^{-1} = -A = A^T$ and $A^2 = -1$.

Consider a transformation of coordinates on the phase space

$$\boldsymbol{X} \to Y_i(\boldsymbol{X}) \tag{9.20}$$

The Jacobian of such a transformation is $\frac{\partial Y_i}{\partial X_j}$. Let us call it $\Delta_{ij}$. The time rate of $\boldsymbol{X}$ is then given by $\dot{\boldsymbol{Y}} = \Delta\dot{\boldsymbol{X}}$. Noting that

$$\frac{\partial H}{\partial X_i} = \frac{\partial H}{\partial Y_j}\frac{\partial Y_j}{\partial X_i} = \Delta^T(\nabla_Y H) \tag{9.21}$$

we get for $\dot{\boldsymbol{X}}$ the equation

$$\dot{\boldsymbol{Y}} = \Delta\dot{\boldsymbol{X}} = \Delta A\frac{\partial H}{\partial X_i} = \Delta A\Delta^T(\nabla_Y H) \tag{9.22}$$

where we used (9.18).

Formally, a $2n \times 2n$ matrix $A$ is said to be symplectic if it satisfies the relation

$$\Delta A\Delta^T = A \tag{9.23}$$

We then see from (9.22) that the Hamiltonian equations are preserved. When the Jacobian $\Delta$ satisfies (9.23) in relation to the antisymmetric block matrix $A$ we say that the transformations are symplectic or canonical. The transformed coordinates, occupying the top and bottom halves of the column matrix $Y$, in view of the mapping (9.20), can be, respectively, identified as the new generalized coordinates $Q_1, Q_2, ..., Q_n$ and generalized momenta $P_1, P_2, ..., P_n$.

**Example 9.3:**

Determine the constants $\lambda, \mu, \nu$ and $\epsilon$ such that

$$Q_1 = \lambda q_1 + \mu q_2, Q_2 = \nu q_1 + \epsilon q_2, P_1 = 2p_1 + p_2, P_2 = p_2$$

is a canonical transformation.

We can carry out the symplectic test to determine the constants $\lambda, \mu, \nu$ and $\epsilon$. To this end we first write down the relevant $(4 \times 4)$ Jacobian matrix which reads

$$\Delta = \begin{pmatrix} \frac{\partial Q_1}{\partial q_1} & \frac{\partial Q_1}{\partial q_2} & \frac{\partial Q_1}{\partial p_1} & \frac{\partial Q_1}{\partial p_2} \\ \frac{\partial Q_2}{\partial q_1} & \frac{\partial Q_2}{\partial q_2} & \frac{\partial Q_2}{\partial p_1} & \frac{\partial Q_2}{\partial p_2} \\ \frac{\partial P_1}{\partial q_1} & \frac{\partial P_1}{\partial q_2} & \frac{\partial P_1}{\partial p_1} & \frac{\partial P_1}{\partial p_2} \\ \frac{\partial P_2}{\partial q_1} & \frac{\partial P_2}{\partial q_2} & \frac{\partial P_2}{\partial p_1} & \frac{\partial P_2}{\partial p_2} \end{pmatrix}$$

For the given relations it takes the form

$$\Delta = \begin{pmatrix} \lambda & \mu & 0 & 0 \\ \nu & \epsilon & 0 & 0 \\ 0 & 0 & 2 & 1 \\ 0 & 0 & 0 & 1 \end{pmatrix}$$

Now $A$ being

$$A = \begin{pmatrix} 0 & 0 & 1 & 0 \\ 0 & 0 & 0 & 1 \\ -1 & 0 & 0 & 0 \\ 0 & -1 & 0 & 0 \end{pmatrix}$$

we find on multiplying the matrix $A\Delta A$ to be given by

$$A\Delta A = \begin{pmatrix} 0 & 0 & 2\lambda + \mu & \mu \\ 0 & 0 & 2\nu + d & \epsilon \\ -2\lambda - \mu & -2\nu - \epsilon & 0 & 0 \\ -\mu & -\epsilon & 0 & 0 \end{pmatrix} \tag{9.24}$$

For the values $\lambda = -\nu = \frac{1}{2}$, $\mu = 0$ and $\epsilon = 1$ we see that $\Delta A \Delta^T = A$ implying that the transformation is canonical. The result is confirmed by the Poisson bracket test which shows that $\{Q_1, Q_1\} = \{Q_2, Q_2\} = 0$, $\{P_1, P_1\} = \{P_2, P_2\} = 0$ and $\{Q_1, P_1\} = \{Q_2, P_2\} = 1$. The transformed coordinates $Q_1$ and $Q_2$ are $Q_1 = \frac{1}{2}q_1, Q_2 = -\frac{1}{2}q_1 + q_2$.

**Example 9.4:**

Show that canonical transformations form a group.

Apart from the identity transformation which is evident, canonical transformations satisfy the product or composition law in that if $\boldsymbol{X} \to \boldsymbol{Y}$ is a canonical transformation due to the Jacobian of the transformation $\Sigma$ satisfying $\Sigma A \Sigma^T = A$ and $\boldsymbol{Y} \to \boldsymbol{Z}$ is another canonical transformation due to the Jacobian of the transformation $\Lambda$ satisfying $\Lambda A \Lambda^T = A$, the product transformation taking $\boldsymbol{X} \to \boldsymbol{Z}$ with the Jacobian $\Delta$ given by $\Delta \equiv \Sigma\Lambda$ is also canonical because $\Delta$ satisfies $\Delta A \Delta^T = \Sigma\Lambda A \Lambda^T \Sigma^T = \Sigma A \Sigma^T = A$. Moreover, if $\boldsymbol{X} \to \boldsymbol{Y}$ is a canonical transformation then the inverse transformation $\boldsymbol{Y} \to \boldsymbol{X}$ is also canonical. For this the determinant of the Jacobian is required to be nonvanishing so that we can carry out the process of inversion. The inverse transformation exists because of $\det\Sigma = 1$. Finally, the associative property is valid which follows by taking three successive transformations characterized by the Jacobians $\Sigma, \Lambda$ and $\Gamma$ mapping successively $\boldsymbol{X} \to \boldsymbol{Y}$, $\boldsymbol{Y} \to \boldsymbol{Z}$ and $\boldsymbol{Z} \to \boldsymbol{T}$. The three transformations may be combined to be interpreted as a single transformation taking $\boldsymbol{X} \to \boldsymbol{T}$ by identifying it as a two-step process either by $\boldsymbol{X} \to \boldsymbol{Z}$ through $\Sigma\Lambda$ followed by $\boldsymbol{Z} \to \boldsymbol{T}$ through $\Gamma$ or by $\boldsymbol{X} \to \boldsymbol{Y}$ through $\Sigma$ followed by $\boldsymbol{Y} \to \boldsymbol{T}$ through $\Lambda\Gamma$. The reason is due to the associative character of a matrix multiplication: $(\Sigma\Lambda)\Gamma = \Sigma(\Lambda\Gamma)$. Thus, all the four properties of a group hold and hence canonical transformations form a group.

---

## 9.3 Idea of a generating function

That the quantity $(pdq - PdQ)$ is an exact differential can be established by noting

$$\begin{aligned} pdq - PdQ &= pdq - P\left(\frac{\partial Q}{\partial q}dq + \frac{\partial Q}{\partial p}dp\right) \\ &= \left(p - P\frac{\partial Q}{\partial q}\right)dq - P\frac{\partial Q}{\partial p}dp \end{aligned} \tag{9.25}$$

from which the condition for an exact differential, namely,

$$\frac{\partial}{\partial p}\left(p - P\frac{\partial Q}{\partial q}\right) = \frac{\partial}{\partial q}\left(-P\frac{\partial Q}{\partial p}\right) \tag{9.26}$$

is seen to hold because of $\{Q, P\}_{(q,p)} = 1$.

Setting $pdq - PdQ = dG_1$, we call $G_1$ to be the generating function of the transformation $(q, p) \to (Q, P)$. Such a class of generating functions is referred to as the Type I generating function. $G_1$ being a function of $q$ and $Q$ [i.e., $G_1 = G_1(q, Q)$] implies

$$dG_1 = \frac{\partial G_1}{\partial q} dq + \frac{\partial G_1}{\partial Q} dQ \tag{9.27}$$

from which we find on matching with the left-hand side of (9.25)

$$p = \frac{\partial G_1}{\partial q}, \quad P = -\frac{\partial G_1}{\partial Q} \tag{9.28}$$

Consider Example 9.1. We first of all check whether $pdq - PdQ$ is a perfect differential for this problem. For the generating function $G_1$, the independent variables are $q$ and $Q$. Expressing $p$ and $P$ in terms of these, namely, $p = m\omega q \cot Q, \quad P = \frac{1}{2} m\omega q^2 \text{cosec}^2 Q$ we can express

$$pdq - PdQ = d\left(\frac{1}{2} m\omega q^2 \cot\theta\right) \tag{9.29}$$

which is indeed an exact differential. Hence $G_1$ for this problem is $G_1(q, Q) = \frac{1}{2} m\omega q^2 \cot\theta$.

There can be other types of generating functions depending on what combinations of old and new canonical variables we choose. For instance a Type II generating function $G_2$ is a function of $q$ and $P$ and the counterparts of (9.28) are

$$p = \frac{\partial G_2}{\partial q}, \quad Q = \frac{\partial G_2}{\partial P} \tag{9.30}$$

For the Type III generating function $G_3$ which is a function of $p$ and $Q$ the underlying relations are

$$q = -\frac{\partial G_3}{\partial p}, \quad P = -\frac{\partial G_3}{\partial Q} \tag{9.31}$$

Finally, a Type IV generating function $G_4$ depends on the pair $(p, P)$ with the complementary variables $q$ and $Q$ satisfying

$$q = -\frac{\partial G_4}{\partial p}, \quad Q = -\frac{\partial G_4}{\partial P} \tag{9.32}$$

The transformations given in (9.28), (9.30), (9.31) and (9.32) hold for the time-independent canonical transformations. These are *restricted* canonical transformations. In the time-dependent case, which we discuss in the next section, the generating functions possess additionally a function of the variable $t$.

## 9.4 Types of time-dependent canonical transformations

Consider a transformation $(q, p, t) \to (Q, P, t)$ for a general time-dependent situation. Since adding a total differential does not change the essential dynamics of the system, we define a canonical transformation $(q, p, t) \to (Q, P, t)$ as the one guided by the following criterion:

$$\sum_{i=1}^{n} pdq - Hdt = \sum_{i=1}^{n} PdQ - Kdt + dG_1 \tag{9.33}$$

with $G_1 = G_1(q, Q, t)$ to be the time-dependent Type I generating function of the transformation. The time-independent generating function was considered earlier from the exact differentiability of the quantity $(pdq - PdQ)$.

We now turn to four kinds of canonical transformations as induced by the corresponding time-dependent generating functions.

### 9.4.1 Type I canonical transformation

A Type I canonical transformation treats the old and new coordinates $q_i$ and $Q_i$ as independent variables. For $G_1(q_i, Q_i, t)$ we have

$$dG_1 = \sum_{i=1}^{n} \left( \frac{\partial G_1}{\partial q_i} dq_i + \frac{\partial G_1}{\partial Q_i} dQ_i \right) + \frac{\partial G_1}{\partial t} dt \tag{9.34}$$

Putting this form in (9.33) we find on comparing the differentials

$$p_i = \frac{\partial G_1(q_i, Q_i, t)}{\partial q_i}, \quad P_i = -\frac{\partial G_1(q_i, Q_i, t)}{\partial Q_i} \tag{9.35}$$

along with

$$K = H + \frac{\partial G_1}{\partial t} \tag{9.36}$$

The first of (9.35) gives $Q_i$ in terms of $q_i$ and $p_i$ which when substituted in the second equation determines $P_i$. Of course, we assume that the Jacobian of the transformations det $|\frac{\partial^2 G_1}{\partial q_i \partial Q_j}| \neq 0)$. $K$ is the new Hamiltonian.

It is worthwhile to note that the Type I canonical transformation induces the exchange transformation. For instance if we choose $G_1(q, Q, t) = q_i Q_i$ then $p_i$ and $P_i$ turn out to be $Q_i$ and -$q_i$, respectively, while $K = H$ reflecting that the coordinates and momenta are exchanged.

### 9.4.2 Type II canonical transformation

For a Type II canonical transformation the independent variables are $q_i$ and $P_i$ which stand, respectively, for the old coordinates and the new momenta. In this case we express

$$\sum_{i=1}^{n} P_i dQ_i = d\left(\sum_{i=1}^{n} P_i Q_i\right) - \sum_{i=1}^{n} Q_i dP_i \tag{9.37}$$

which results in

$$\sum_{i=1}^{n} p_i dq_i + \sum_{i=1}^{n} Q_i dP_i - H dt = d\left(\sum_{i=1}^{n} P_i Q_i\right) - K dt + dG_1 \tag{9.38}$$

The accompanying generating function $G_2$ can therefore be defined by

$$G_2 = G_1 + \sum_{i=1}^{n} P_i Q_i \tag{9.39}$$

Viewing $G_2$ as a function of $q_i, P_i$ and $t$ we write

$$dG_2(q_i, P_i, t) = \sum_{i=1}^{n}\left(\frac{\partial G_2}{\partial q_i}dq_i + \frac{\partial G_2}{\partial P_i}dP_i\right) + \frac{\partial G_2}{\partial t}dt \tag{9.40}$$

which implies from (9.38) and (9.39)

$$p_i = \frac{\partial G_2(q_i, P_i, t)}{\partial q_i}, \quad Q_i = \frac{\partial G_2(q_i, P_i, t)}{\partial P_i} \tag{9.41}$$

along with

$$K = H + \frac{\partial G_2}{\partial t} \tag{9.42}$$

The first of (9.41) gives $P_i$ in terms of $q_i$ and $p_i$ which when substituted in the second equation provides for $Q_i$. The Jacobian of the transformation is assumed to be nonvanishing: $\det|\frac{\partial^2 G_2}{\partial q_i \partial P_j}| \neq 0$ . $K$ as given by (9.42) is the new Hamiltonian.

The Type II canonical transformation has in its embedding both the identity as well as the point transformations. In fact, corresponding to $G_2(q, P, t) = q_i P_i$ we see that $p_i = P_i$, $Q_i = q_i$ and $K = H$ showing for the identity transformation while for $G_2(q, P, t) = \phi(q_i, t)P_i$ we have $Q_i = \phi(q_i, t)$ implying that the new coordinates are functions of old coordinates.

### 9.4.3 Type III canonical transformation

In Type III canonical transformation the underlying generating function $G_3$ is a function of the independent variables $p_i, Q_i$ which are, respectively, the old momenta and new coordinates. $G_3$ is defined by

$$G_3(p_i, Q_i, t) = G_1 - \sum_{i=1}^{n} q_i p_i \tag{9.43}$$

which leads to

$$q_i = -\frac{\partial G_3}{\partial p_i}, \quad P_i = -\frac{\partial G_3}{\partial Q_i}, \quad K = H + \frac{\partial G_3}{\partial t} \tag{9.44}$$

Here the Jacobian of the transformation is assumed to be nonvanishing: $\det |\frac{\partial^2 G_3}{\partial p_i \partial Q_j}| \neq 0$.

### 9.4.4 Type IV canonical transformation

In the Type IV canonical transformation the generating function is a function of old and new momenta and given by

$$G_4(p_i, P_i, t) = G_1 + \sum_{i=1}^{n} Q_i P_i - \sum_{i=1}^{n} q_i p_i \tag{9.45}$$

in which $p_i$ and $P_i$ are treated as independent variables. We then have

$$q_i = -\frac{\partial G_4}{\partial p_i}, \quad Q_i = \frac{\partial G_4}{\partial P_i}, \quad K = H + \frac{\partial G_4}{\partial t} \tag{9.46}$$

Here the Jacobian of the transformation is assumed to be nonvanishing: $\det |\frac{\partial^2 G_4}{\partial p_i \partial P_j}| \neq 0$.

**Example 9.5**

Consider the transformation

$$Q = -p, \quad P = q + \lambda p^2$$

where $\lambda$ is a constant. By the Poisson bracket test, namely, $[Q, P]_{(q,p)} = 1$ we conclude that the transformation is a canonical transformation.

The Type 1 generating function is determined by showing $pdq - PdQ$ to be a perfect differential:

$$\begin{aligned} pdq - PdQ &= (-Q)dq - (q + \lambda Q^2)dQ \\ &= d(-qQ - \frac{1}{3}\lambda Q^3) \end{aligned}$$

Thus

$$G_1(q,Q) = -qQ - \frac{1}{3}\lambda Q^3$$

On the other hand, the Type 2 generating function is obtained by treating $q$ and $P$ as independent variables. From (9.39) we have

$$\begin{aligned} G_2(q,p) &= G_1 + PQ \\ &= -qQ - \frac{1}{3}\lambda Q^3 + (q+\lambda Q^2)Q \\ &= \frac{2}{3}\lambda Q^3 \\ &= \frac{2}{3}\lambda\left(\frac{P-q}{\lambda}\right)^{\frac{3}{2}} \end{aligned}$$

Check that

$$\begin{aligned} \left(\frac{\partial G_2}{\partial q}\right)_P &= -\frac{1}{\sqrt{\lambda}}(P-q)^{\frac{1}{2}} = p \\ \left(\frac{\partial G_2}{\partial P}\right)_q &= \frac{1}{\sqrt{\lambda}}(P-q)^{\frac{1}{2}} = Q \end{aligned}$$

which are as required.

**Example 9.6**

The transformations

$$\begin{aligned} Q &= q\cos\theta - \frac{p}{m\omega}\sin\theta \\ P &= m\omega q\sin\theta + p\cos\theta \end{aligned}$$

are easily seen to be canonical due to $\{Q,P\}_{(q,p)} = 1$. We also find

$$\begin{aligned} p &= m\omega\left(q\cot\theta - \frac{Q}{\sin\theta}\right) \\ P &= m\omega\left(\frac{q}{\sin\theta} - Q\cot\theta\right) \end{aligned}$$

Hence the quantity $pdq - PdQ$ can be expressed as

$$pdq - PdQ = d\left[\frac{1}{2}m\omega(q^2+Q^2)\cot\theta - m\omega qQ\text{cosec}\theta\right]$$

The Type 1 generating function $G_1(q,Q)$ is identified as

$$G_1(q,Q) = \frac{1}{2}m\omega(q^2+Q^2)\cot\theta - m\omega qQ\text{cosec}\theta$$

On other other hand, the Type 2 generating function can be obtained from $G_2 = G_1 + PQ$. We get

$$G_2(q,Q) = \frac{1}{2}m\omega(q^2 - Q^2)\cot\theta$$

To assign the right variable dependence on $G_2$, namely, $q$ and $P$ we note that

$$Q = \frac{q}{\cos\theta} - \frac{P}{m\omega}\tan\theta$$

by eliminating $p$ from the given transformations. Substituting for $Q$, $G_2$ turns out to be

$$G_2(q,P) = \frac{qP}{\cos\theta} - \frac{1}{2}m\omega\left(q^2 + \frac{P^2}{m^2\omega^2}\right)\tan\theta$$

We can verify that

$$\begin{aligned}
\left(\frac{\partial F_2}{\partial q}\right)_P &= \frac{P}{\cos\theta} - m\omega q\tan\theta = p \\
\left(\frac{\partial F_2}{\partial P}\right)_q &= \frac{q}{\cos\theta} - \frac{P}{m\omega}\tan\theta = Q
\end{aligned}$$

**Example 9.7**

We consider the harmonic oscillator Hamiltonian $H$ which is invariant under the set of canonical transformations considered in the previous example:

$$\begin{aligned}
H(q,p) &= \frac{1}{2m}p^2 + \frac{1}{2}m\omega^2 q^2 \\
\to &\quad \frac{1}{2m}P^2 + \frac{1}{2}m\omega^2 Q^2 = H(Q,P)
\end{aligned}$$

The generating function $G_2(q,P)$ which was found to be

$$G_2(q,P) = \frac{qP}{\cos\theta} - \frac{1}{2}m\omega\left(q^2 + \frac{P^2}{m^2\omega^2}\right)\tan\theta$$

gives for a partial derivative with respect to time

$$\begin{aligned}
\left(\frac{\partial G_2}{\partial t}\right)_{q,P} &= \left[qP\sin\theta - \frac{1}{2}m\omega\left(q^2 + \frac{p^2}{m^2\omega^2}\right)\right]\sec^2\theta\dot{\theta} \\
&= -\left(\frac{P^2}{2m\omega} + \frac{1}{2}m\omega Q^2\right)\dot{\theta} \\
&= -H(Q,P)\frac{\dot{\theta}}{\omega}
\end{aligned}$$

Therefore, the transformed Hamiltonian reads

$$K(Q,P,t) = \left(1 - \frac{\dot{\theta}}{\omega}\right) H(Q,P)$$

which vanishes for $\dot{\theta} = \omega$, i.e., $\theta = \omega t$. As a consequence $Q = \frac{\partial K}{\partial P} = 0$ and $\dot{P} = -\frac{\partial K}{\partial Q} = 0$. Hence $Q$ and $P$ are constants which we write as

$$Q = Q_0, \quad P = P_0$$

where $Q_0$ and $P_0$ stand for the initial values of $q$ and $p$, respectively.

Reverting to the old coordinates we get

$$\begin{aligned} q(t) &= Q_0 \cos\omega t + \frac{P_0}{m\omega} \sin\omega t \\ p(t) &= -m\omega Q_0 \sin\omega t + P_0 \cos\omega t \end{aligned}$$

which give the time evolution for $q$ and $p$.

---

## 9.5 Infinitesimal canonical transformations

We remarked earlier that the Type II canonical transformation is well suited for the identity transformation. An infinitesimal canonical transformation $(q,p) \to (Q,P)$ about the identity is generated through an expression of the form

$$G_2(q,P,t) = q_i P_i + \epsilon G(q,P) \tag{9.47}$$

where $\epsilon$ is an infinitesimal quantity and results in

$$p_i = \frac{\partial G_2(q_i,P_i,t)}{\partial q_i} = P_i + \epsilon \frac{\partial G}{\partial q_i}, \quad Q_i = \frac{\partial G_2(q_i,P_i,t)}{\partial P_i} = q_i + \epsilon \frac{\partial G}{\partial P_i} \tag{9.48}$$

with $K = H$. The departures from the exact identity are due to the presence of small quantities $\epsilon \frac{\partial G}{\partial q_i}$ and $\epsilon \frac{\partial G}{\partial P_i}$ in $p_i$ and $Q_i$, respectively. We can summarize $\delta q_i \equiv Q_i - q_i$ and $\delta p_i \equiv P_i - p_i$ as

$$\delta q_i = \epsilon \frac{\partial G(q,P)}{\partial P_i} = \epsilon \frac{\partial G(q,p)}{\partial p_i} + O(\epsilon^2), \ \delta p_i = -\epsilon \frac{\partial G(q,P)}{\partial q_i} = -\epsilon \frac{\partial G(q,p)}{\partial q_i} + O(\epsilon^2) \tag{9.49}$$

As an immediate role of the above results we can think of the Hamiltonian flow following the chain given by $(q(t), p(t)) \to (Q(t), P(t)) \equiv (q(t+\delta t), p(t+\delta t))$:

$$Q_i(t) = q_i(t) + \dot{q}_i \delta t = q_i(t) + \frac{\partial H}{\partial p_i}\delta t, \quad P_i(t) = p_i(t) + \dot{p}_i \delta t = p_i(t) - \frac{\partial H}{\partial q_i}\delta t \tag{9.50}$$

where we Taylor expanded $q_i(t+\delta t)$ and $p_i(t+\delta t)$ up to $O(\delta t)$.

$\delta q_i$ and $\delta p_i$ implied by (9.49) when compared with (9.50) leads to the identification of the Hamiltonian $H$ with the generating function $G$ of the infinitesimal canonical transformation, i.e., $H = G$ while the increment $\delta t$ with the small quantity $\epsilon$, i.e., $\delta t = \epsilon$. Thus, the Hamiltonian can be looked upon as the generator of the infinitesimal transformation and the time evolution as a one-parameter canonical transformation.

---

## 9.6 Hamilton–Jacobi equation

The idea of deriving the Hamilton–Jacobi equation rests on effecting a canonical transformation that maps one basis of known canonical variables to a new set of coordinates and momenta such that the latter are cyclic for the transformed Hamiltonian. What does it mean?

Consider the specific case of a Type II generating function $G_2(q_i, P_i, t), i = 1, 2, ..., n$ which facilitates the transformations according to (9.41) where the time derivative of $Q_i$ and $P_i$ stand as

$$\dot{Q}_i = \frac{\partial K}{\partial P_i}, \quad \dot{P}_i = -\frac{\partial K}{\partial Q_i}, \quad i = 1, 2, ..., n$$

as defined by (9.2). Now if $Q_i$ and $P_i$ are cyclic coordinates in the transformed Hamiltonian $K$, then it follows that these have to be constants in time. Such a validity is assured if we assume, without loss of generality, $K$ to be zero. It then gives from (9.42) the equation

$$H(q_1, q_2, ..., q_n; p_1, p_2, ..., p_n; t) + \frac{\partial G_2}{\partial t} = 0$$

A few remarks are in order:

From (9.40) a total derivative of the generating function $G_2(q_i, p_i, t)$ with respect to $t$ gives, in view of the above equation,

$$\frac{dG_2}{dt} = \sum_{i=1}^{n} \frac{\partial G_2}{\partial q_i}\dot{q}_i = \sum_{i=1}^{n} p_i \dot{q}_i - H(q_i, p_i, t) = L \tag{9.51}$$

where we employed (9.41) and noted that $P_i$'s are cyclic coordinates in $K$ and hence taken to be constants. Thus, integrating between times, say, $t_1$ and

$t_2$ the quantity $S \equiv G_2$ defined by the action integral $S = \int_{t_1}^{t_2} Ldt$ satisfies (9.41).

Replacing $p_i$ by the corresponding relation given in the first of (9.41) we arrive at the form

$$p_i = \frac{\partial S}{\partial q_i}: \quad H(q_1, q_2, ..., q_n; \frac{\partial S}{\partial q_1}, \frac{\partial S}{\partial q_2}, ..., \frac{\partial S}{\partial q_n}; t) + \frac{\partial S}{\partial t} = 0 \tag{9.52}$$

which is called the time-dependent Hamilton–Jacobi equation. More compactly it is also expressed as

$$H(q_i, \frac{\partial S}{\partial q_i}, t) + \frac{\partial S}{\partial t} = 0 \tag{9.53}$$

A function $S$ which is a solution of the above is called the Hamilton principal function.

Equation (9.53) is a first-order differential equation involving the $n$ coordinates $q_i$'s and $t$. It is not linear in $\frac{\partial S}{\partial q_i}$ because the partial derivatives of $S$ appear in higher degree than the first. Associated with the $(n+1)$ variables ($n$ $q_i$'s and $t$) we expect $(n+1)$ constants of motion namely $\alpha_1, \alpha_2, ..., \alpha_n, \alpha_{n+1}$. However, we notice one curious thing in (9.53) which is that the dependent variable $S$ itself does not appear in it: only its partial derivatives do. So one of the constants has no bearing on the solution, i.e., the solution has an additive constant. Disregarding such an irrelevant additive constant we write for the complete integral of (9.54) the form

$$S = S(q_1, q_2, ..., q_n; \alpha_1, \alpha_2, ..., \alpha_n; t) \tag{9.54}$$

where it is ensured that none of the constants $\alpha$'s is of additive nature to the solution. Actually we identify these constants with the constant momenta $P_i$, $i = 1.2..., n$, which, according to our choice, were assumed to be cyclic.

The cyclic coordinates $Q_i$'s provide another set of constants $\beta_i' s, i = 1, 2, ..., n$ and read from the second equation of (9.41)

$$Q_i = \frac{\partial S}{\partial \alpha_i} = \beta_i, \quad i = 1, 2, ..., n \tag{9.55}$$

We clarify the above issues by considering two examples: one for the free particle and the other that of the harmonic oscillator.

**Example 9.8**

The Hamilton–Jacobi equation for the free particle problem is obviously

$$\frac{1}{2m}(\frac{\partial S}{\partial q})^2 + \frac{\partial S}{\partial t} = 0$$

Its complete integral can be easily ascertained by inspection which reads

$$S(q, \alpha, t) = \sqrt{2m\alpha}q - \alpha t$$

where $\alpha$ is a non-additive constant.

From (9.55)

$$\beta = \frac{\partial S}{\partial \alpha} = k \quad \text{(say)}$$

$$\text{i.e.} \qquad k = \sqrt{\frac{m}{2\alpha}}q - t \quad \Rightarrow \quad q = \sqrt{\frac{2E}{m}}(t + k) \quad \text{and } p = \frac{\partial S}{\partial q} = \sqrt{2m\alpha}.$$

which conform to their expected forms.

It is to be noted that Hamilton–Jacobi equation being a partial differential equation can admit multiple solutions. For instance, in the above case, there is also a legitimate solution given by

$$S(q, \alpha, t) = \frac{m(q - \alpha)^2}{2t}$$

where $\alpha$ is a non-additive constant. This implies

$$\beta = \frac{\partial S}{\partial \alpha} = -\frac{m}{t}(q - \alpha)$$

giving the time-dependence: $q(t) = \alpha - \frac{\beta}{m}t$ where $\alpha$ can be identified as the initial value of $q$ while $\beta$ is the negative of the momentum: $\beta = -p$.

**Example 9.9**

For the harmonic oscillator problem the Hamilton–Jacobi equation has the form

$$H = \frac{1}{2m}\left(p^2 + m^2\omega^2 q^2\right)$$

The relation $p = \frac{\partial S}{\partial q}$ gives

$$\frac{1}{2m}\left[\left(\frac{\partial S}{\partial q}\right)^2 + m^2\omega^2 q^2\right] + \frac{\partial S}{\partial t} = 0$$

To solve for the above equation we try separation of variables for $S$:

$$S(q : \alpha; t) = W(q : \alpha) - \alpha t$$

where $\alpha$ is a non-additive constant.The function $W(q{:}\alpha)$ is the time-independent part called Hamilton's characteristic function which we shall discuss in the next section.

We get

$$\frac{1}{2m}\left[\left(\frac{\partial W}{\partial q}\right)^2+m^2\omega^2q^2\right]=\alpha$$

which gives the integral

$$W=\sqrt{2m\alpha}\int dq\sqrt{1-\frac{m\omega^2q^2}{2\alpha}}$$

Such an integral implies for $\beta$

$$\beta=\frac{\partial S}{\partial\alpha}=\sqrt{\frac{m}{2\alpha}}\int\frac{dq}{\sqrt{1-\frac{m\omega^2q^2}{2\alpha}}}-t$$

which integrates to

$$t+\beta=\frac{1}{\omega}\sin^{-1}q\sqrt{\frac{m\omega^2}{2\alpha}}$$

Hence the solutions for $q$ and $p$ are

$$\begin{aligned}q&=\sqrt{\frac{2\alpha}{m\omega^2}}\sin\omega(t+\beta)\\ p&=\frac{\partial S}{\partial q}=\frac{\partial W}{\partial q}=\sqrt{2m\alpha-m\omega^2q^2}=\sqrt{2m\alpha}\cos\omega(t+\beta)\end{aligned}$$

which are in agreement with their well-known forms.

### 9.6.1 Time independent Hamilton–Jacobi equation: Hamilton's characteristic function

In the time independent case the Hamilton does not depend on time and the system is conservative. The Hamilton–Jacobi equation reduces to the form

$$H\left(q_i,\frac{\partial S}{\partial q_i}\right)+\frac{\partial S}{\partial t}=0 \tag{9.56}$$

where $p_i=\frac{\partial S}{\partial q_i}$. Note that the time dependent character of $S$ stays. A solution of (9.56) is given in the separation of variables form

$$S(q_i,\alpha_i,t)=W(q_i,\alpha_i)-Et \tag{9.57}$$

implying

$$p_i=\frac{\partial W}{\partial q_i}:\quad H\left(q_i,\frac{\partial S}{\partial q_i}\right)=H\left(q_i,\frac{\partial W}{\partial q_i}\right)=E \tag{9.58}$$

where the Hamiltonian has the constant value $E$. $W$ is called Hamilton's characteristic function. Of the two equations in (9.58) the first one is the time-independent Hamilton–Jacobi equation while the second one is the equation for Hamilton's characteristic function.

The separability of the equation for Hamilton's characteristic function means we should be able to express $W$ as

$$W = W_1(q_1;\alpha_1,\alpha_2,...,\alpha_n)+W_2(q_2;\alpha_1,\alpha_2,...,\alpha_n)+...+W_n(q_n;\alpha_1,\alpha_2,...,\alpha_n) \tag{9.59}$$

This has the effect of projecting the characteristic function equation as

$$H_i(q_i,\frac{\partial W_i(q_i;\alpha_1,\alpha_2,...,\alpha_n)}{\partial q_i}) = E, \quad i=1,2,...,n \tag{9.60}$$

In general each $W_i$ will contain the corresponding $q_i$ along with the set of $\alpha$'s but there may be situations when one $W_i$ will contain one $q_i$ and one $\alpha_i$ for $i=1,2,...,n$ as the following example demonstrates.

Consider a system with one degree of freedom having the Hamiltonian

$$H = \frac{p^2}{2m} + V(q) \tag{9.61}$$

It results in the Hamilton–Jacobi equation

$$\frac{1}{2m}(\frac{\partial S}{\partial q})^2 + V(q) + \frac{\partial S}{\partial t} = 0 \tag{9.62}$$

Employing separation of variables we express

$$S(q,\alpha,t) = W(q,\alpha) - Et \tag{9.63}$$

where the separation constant $E$ is identified as the obvious non-additive constant. It leads to an ordinary differential equation for $W$

$$\frac{1}{2m}(\frac{dW}{dq})^2 + V(q) = E \tag{9.64}$$

which can be inverted to read in the integral form

$$W = \int \sqrt{2m(E-V(q))}dq \tag{9.65}$$

We do not need to determine $W$ explicitly. What suffices is the knowledge for the quantities $q$ and $\beta$ which are obtainable from

$$p = \frac{\partial W}{\partial q} = \sqrt{2m(\alpha - V(q))}, \quad \beta + t = \frac{\partial W}{\partial E} = \sqrt{\frac{m}{2}}\int \frac{dx}{\sqrt{(2m(E-V(q)))}} \tag{9.66}$$

We give below some illustrations how such a scheme works in practice.

**Example 9.10**

We again focus on the harmonic oscillator potential $V(q) = \frac{1}{2}m\omega^2 q^2$. Putting $q = \sqrt{\frac{2E}{m\omega^2}}\sin\phi$, an integration gives from (9.65)

$$W = \frac{E}{\omega}(\phi + \sin\phi\cos\phi)$$

Further from (9.66) we obtain

$$p = \frac{\partial W}{\partial q} = \sqrt{2mE}\cos\phi$$

$$\beta + t = \frac{\partial W}{\partial E} = \frac{\phi}{\omega}$$

**Example 9.11**

Consider the vertical motion of a particle in a unform gravitational field as given by the potential $V(z) = mqz$ where we have identified the generalized coordinate $q$ with $z$. The Hamiltonian is

$$H = \frac{p^2}{2m} + mgz = E = \alpha$$

which for $p = \frac{\partial W}{\partial z}$ gives the differential equation

$$\frac{1}{2m}(\frac{dW}{dz})^2 + mgz = \alpha$$

Exploiting the second equation of (9.66) gives

$$g(\beta + t) = \sqrt{\frac{2}{m}}\sqrt{E - mgz}$$

which on squaring leads to

$$z = -\frac{1}{2}g(\beta + t)^2 + \frac{E}{mg}$$

On the other hand, for $p$ we have from the first equation of (9.66)

$$p = \frac{\partial W}{\partial z} = mg(\beta + t)$$

If we use initial conditions that at $t = 0$, $z = z_0, p = 0$ then we have $E = mgz_0$ and $\beta = 0$ furnishing the standard expressions $z = -\frac{1}{2}gt^2 + z_0$ and $p = mgt$.

### 9.6.2 Other variants of Hamilton–Jacobi equation

We derived the Hamilton–Jacobi equation by exploiting a Type II generating function. But we could as well have used the other types. Here is a summary of Hamilton–Jacobi equations resulting from Type I, Type III and Type IV generating functions:

$$\text{Type I}: H(q_i, p_i, t) + \frac{\partial G_1(q, Q, t)}{\partial t} = 0 \Rrightarrow H(q_i, \frac{\partial S}{\partial q_i}, t) + \frac{\partial S(q_i, Q_i, t)}{\partial t} = 0 \tag{9.67}$$

$$\text{Type III}: H(q_i, p_i, t) + \frac{\partial G_3(p, Q, t)}{\partial t} = 0 \Rrightarrow H(-\frac{\partial S}{\partial p_i}, p_i, , t) + \frac{\partial S(p_i, Q_i, t)}{\partial t} = 0 \tag{9.68}$$

$$\text{Type IV}: H(q_i, p_i, t) + \frac{\partial G_4(p_i, P_i, t)}{\partial t} = 0 \Rrightarrow H(-\frac{\partial S}{\partial p_i}, p_i, , t) + \frac{\partial S(p_i, P_i, t)}{\partial t} = 0 \tag{9.69}$$

Note that in both (9.68) and (9.69) the coordinates in the Hamilton–Jacobi equations appear as partial derivatives which are rather complicated to handle. So these do not have much utility. The form (9.67) is relatively simpler compared to (9.68) and (9.69) and is sometimes used but one has to remember that the Type I and Type II generating functions are related as given in (9.39).

---

## 9.7 Action-angle variables

We are familiar with ideas of finite motions that are periodic. For a system with one degree of freedom implying that the phase space is of dimension two, if the motion is described by a closed curve, then both $q$ and $p$ variables return to the same point periodically in the manner

$$q(t + \tau) = q(t), \quad p(t + \tau) = p(t) \tag{9.70}$$

where $\tau$ is a constant quantity. We call the motion liberation or oscillatory in character.

It can sometimes happen that while the $p$-variable is periodic, the periodicity of the $q$ variable is subject to a displacement by an amount $q_0$ after each period $\tau$:

$$p(t + \tau) = p(t), \quad q(t + \tau) = q(t) + q_0 \tag{9.71}$$

The rotation has such a feature.

Let us consider a canonical transformation as guided by $S$ in terms of the characteristic function $W(q_i, \alpha_i)$. Let us define the quantities $J_i$ by

$$J_I = \oint p_i dq_i, \quad i = 1, 2, ..., n \tag{9.72}$$

where for the integral we take the coordinate $q_i$ around a closed path while other coordinates remain spectators. Replacing $p_i$ by the partial derivative $\frac{\partial W_i(q_i;\alpha)}{\partial q_i}$, we can express $J_i$ as

$$J_i = \oint \frac{\partial W_i(q_i;\alpha)}{\partial q_i} dq_i, \quad i = 1, 2, ..., n \tag{9.73}$$

Since $W_i$ depends on the variables $q_i$'s and $\alpha$'s and $q_i$'s are integrated over, $J_i$'s finally become only a function of the constants $\alpha_1, \alpha_2, ..., \alpha_n$:

$$J_i = J_i(\alpha_1, \alpha_2, ..., \alpha_n), \quad i = 1, 2, ..., n \tag{9.74}$$

$J_i$ is called the action variable for the $i^{th}$ degree of freedom.

Inverting (9.74) it follows that

$$\alpha_i = \alpha_i(J_1, J_2, ..., J_n), \quad i = 1, 2, ..., n \tag{9.75}$$

We can thus write from (9.59)

$$\begin{aligned} W(q_i, J_i) &= W_1(q_1; J_1, J_2, ..., J_n) + W_2(q_2; J_1, J_2, ..., J_n) + ... \\ &+ W_1(q_n; J_1, J_2, ..., J_n) \end{aligned} \tag{9.76}$$

The coordinate $\theta_i$ which is conjugate to $J_i$ is called the angle variable. This along with $p_i$ are given by

$$\theta_i = \frac{\partial W_i(q_i, J_i)}{\partial J_i}, \quad p_i = \frac{\partial W}{\partial q_i}, \quad i = 1, 2 =, ..., n \tag{9.77}$$

The dimension of $J_i$ is that of the "action" and $W(q_i, J_i)$ is, in effect, the generating function of the canonical transformation $(q_i, p_i) \rightarrow (\theta_i, J_i)$.

The time independence of $W$ makes the old and transformed Hamiltonians to be the same. The Hamiltonian which is only a function of the constants $\alpha_1, \alpha_2, ..., \alpha_n$ is now a function of action variables $J_1, J_2, ..., J_n$ because of (9.75). The time derivatives of $J_i$ and $\theta_i$ read

$$\dot{J}_i = -\frac{\partial H}{\partial \theta_i} = 0, \quad \dot{\theta}_i = \frac{\partial H}{\partial J_i} = \theta_i(J_i), \quad i = 1, 2 =, ..., n \tag{9.78}$$

An integration reveals that $J_i$'s are constants while $\theta_i$'s are linear functions of time given by $\theta_i = \theta_i(J_i)t + \text{constant}$, involving the constants $\omega_i$'s which in turn depend upon $J_i$'s.

Consider the pair $(J, \theta)$. If now $\theta$ is chosen in such a way that it increases by $2\pi$ as the motion undergoes one period, the time period will be given by

$T = \frac{2\pi}{\theta(J)}$ where we call $\theta(J)$ to be the angular frequency. The phase curve will be characterized by

$$q = q(\theta + 2\pi, J) = q(\theta, J), \quad p = p(\theta + 2\pi, J) = p(\theta, J) \tag{9.79}$$

reflecting the $2\pi$-periodic dependence on $\theta$ of $p$ and $q$ variables.

**Example 9.12**

Let us take the harmonic oscillator problem for which the Hamiltonian is $H = \frac{1}{2m}p^2 + \frac{1}{2}m\omega^2 q^2$ with $H = E = \alpha$. The Hamilton's characteristic function obeys

$$\frac{1}{2m}(\frac{dW}{dq})^2 + \frac{1}{2}m\omega^2 q^2 = \alpha$$

From (9.65) $W$ can be put in the form

$$W = m\omega \int \sqrt{\frac{2\alpha}{m\omega^2} - q^2 dq}$$

Using it and the definition for the action variable

$$J = \oint \frac{\partial W(q;\alpha)}{\partial q} dq \tag{9.80}$$

it is a straightforward task to perform the integration by substituting $q = \sqrt{\frac{2\alpha}{m\Omega^2}} \sin\theta$ and taking $(0, 2\pi)$ as the range for $\theta$. It gives for $J$

$$J = 2\pi \frac{\alpha}{\omega}$$

As such we can express

$$H = \alpha = \frac{J\omega}{2\pi}$$

Taking partial derivative of $H$ with respect to $J$ we then find for the angle variable

$$\theta = \frac{\omega}{2\pi}$$

Note that $\omega$ being the angular frequency, the angle variable coincides with the ordinary frequency of the oscillator.

**Example 9.13**

The potential of a two-dimensional oscillator with different angular frequencies $\Omega_x$ and $\Omega_y$ is given by

$$V(x, y) = \frac{1}{2}m\Omega_x^2 x^2 + \frac{1}{2}m\Omega_y^2 y^2$$

The equation for the characteristic function reads

$$\frac{1}{2m}\left[\left(\frac{\partial W}{\partial x}\right)^2+\left(\frac{\partial W}{\partial y}\right)^2\right]+\frac{1}{2}m\Omega_x^2x^2+\frac{1}{2}m\Omega_y^2y^2=E=\alpha_x+\alpha_y \text{ (say)}$$

Adopting a separation of variables for the $x$ and $y$ coordinates for $W$, i.e., writing $W = X(x) + Y(y)$, we obtain straightforwardly from the above equation

$$W=\int\sqrt{2m\alpha_x-m^2\Omega_x^2x^2}dx+\int\sqrt{2m\alpha_y-m^2\Omega_y^2y^2}dy$$

In consequence the two momenta are

$$p_x=\frac{\partial W}{\partial x}=\sqrt{2m\alpha_x-m^2\Omega_x^2x^2},\quad p_y=\frac{\partial W}{\partial y}=\sqrt{2m\alpha_y-m^2\Omega_y^2y^2}$$

The corresponding action variables are

$$J_x=\oint\sqrt{2m\alpha_x-m^2\Omega_x^2x^2}dx,\quad J_y=\oint\sqrt{2m\alpha_y-m^2\Omega_y^2y^2}dy$$

Substituting $x=\sqrt{\frac{2m\alpha_x}{m^2\Omega_x^2}}\sin\phi,\quad y=\sqrt{\frac{2m\alpha_y}{m^2\Omega_y^2}}\sin\theta$ we obtain for $J_x$ and $J_y$ when integrated in the range $(0,2\pi)$ the following forms:

$$J_x=\frac{2\pi\alpha_x}{\Omega_x},\quad J_y=\frac{2\pi\alpha_y}{\Omega_y}$$

These imply

$$\alpha_x=\frac{\Omega_xJ_x}{2\pi},\quad \alpha_y=\frac{\Omega_yJ_y}{2\pi}$$

Next, the angle variables being $\phi_x=\frac{\partial W}{\partial J_x}$ and $\phi_y=\frac{\partial W}{\partial J_y}$ can be evaluated to be $\frac{\Omega_x}{2\pi}$ and $\frac{\Omega_y}{2\pi}$, respectively, and correspond to the ordinary frequencies in the $x$ and $y$ directions.

**Example 9.14**

Consider the Keplerian motion of a particle in a central field of force. The kinetic and potential energies are

$$T=\frac{1}{2}m(\dot{r}^2+r^2\dot{\theta}^2+r^2\sin^2\theta\dot{\phi}^2),\quad V(r)=-\frac{\mu}{r}$$

where $\mu$ is a real constant. Then the Hamiltonian is given by

$$H=\frac{1}{2m}\left(p_r^2+\frac{p_\theta^2}{r^2}+\frac{p_\phi^2}{r^2\sin^2\theta}\right)-\frac{\mu}{r}$$

where we have employed the Legendre transformation $H = \sum_{i=1}^{n} p_i\dot{q}_i - L$ with $L = T - V$. The conjugate momenta are $p_r = m\dot{r}, p_\theta = mr^2\dot{\theta}$ and $p_\phi = mr^2 \sin^2\theta\dot{\phi}$.

To obtain the Hamilton–Jacobi equation we make the replacements

$$p_r = \frac{\partial S}{\partial r}, \quad p_\theta = \frac{\partial S}{\partial \theta} \quad p_\phi = \frac{\partial S}{\partial \phi}$$

yielding for $S$ the partial differential equation

$$\frac{1}{2m}[(\frac{\partial S}{\partial r})^2 + \frac{1}{r^2}(\frac{\partial S}{\partial \theta})^2 + \frac{1}{r^2\sin^2\theta}(\frac{\partial S}{\partial \phi})^2] - \frac{\mu}{r} + \frac{\partial S}{\partial t} = 0$$

Taking for S the form

$$S(r,\theta,\phi,\alpha,t) = W(r,\theta,\phi,\alpha) - \alpha t$$

results in the corresponding equation for the characteristic function $W$

$$\frac{1}{2m}[(\frac{\partial W}{\partial r})^2 + \frac{1}{r^2}(\frac{\partial W}{\partial \theta})^2 + \frac{1}{r^2\sin^2\theta}(\frac{\partial W}{\partial \phi})^2] - \frac{\mu}{r} = \alpha = -|E|$$

where we have taken the energy $-|E| < 0$ for a bound state.

Assuming separation of variables for the function $W$ given by $W = W_r(r) + W_\theta(\theta) + W_\phi(\phi)$, the above equation can be put in the form

$$(\frac{dW_r}{dr})^2 + \frac{1}{r^2}(\frac{dW_\theta}{d\theta})^2 + \frac{1}{r^2\sin^2\theta}(\frac{dW_\phi}{d\phi})^2 = 2m(-|E| + \frac{\mu}{r})$$

where the partial derivatives have been replaced by ordinary derivatives.

By multiplying with $r^2\sin^2\theta$, the only $\phi$-dependent term can be isolated enabling us to express

$$\frac{dW_\phi}{d\phi} = \alpha_\phi = p_\phi$$

where $\alpha_\phi$ is a constant.

Similarly carrying out separation of variables for the other two variables $r$ and $\theta$, $p_\theta$ and $p_r$ result from solving the corresponding differential equations. These are

$$\frac{dW_\theta}{d\theta} = p_\theta = \sqrt{\alpha_\theta^2 - \alpha_\phi^2 \mathrm{cosec}^2\theta}$$

and

$$\frac{dW_r}{dr} = p_r = \sqrt{2m(-|E| + \frac{\mu}{r}) - \frac{\alpha_\theta^2}{r^2}}$$

and $W$ can be obtained from the integral

$$W = \int (p_r dr + p_\theta d\theta + p_\phi d\phi)$$

Of the three action variables, $J_r$, $J_\theta$ and $J_\phi$, the last one $J_\phi$ is the easiest to evaluate. We straightforwardly find

$$J_\phi = \oint p_\phi d\phi = \oint \frac{dW_\phi}{d\phi} = \oint \alpha_\phi d\phi = 2\pi\alpha_\phi$$

Determination of $J_r$ and $J_\theta$ is relatively more complicated. First consider $J_r$. It is given by

$$J_r = \oint p_r dr = \oint \frac{dW_r}{dr} = \oint \sqrt{2m(-|E| + \frac{\mu}{r}) - \frac{\alpha_\theta^2}{r^2}} dr$$

If we take a partial derivative of $J_r$ through the integral we can write it in the way[1]

$$\frac{\partial J_r}{\partial r} = -\sqrt{\frac{2m}{|E|}} \int_a^b \frac{dr}{\sqrt{(a-r)(r-b)}}$$

where $a$ and $b$ are the turning points where $p_r$ vanishes. From the form for $p_r$ given above, we solve for the quadratic in $r$ to obtain for $a$ and $b$

$$a + b = \frac{\mu}{|E|}, \quad ab = \frac{\alpha_\theta^2}{2m|E|}$$

on comparing with the integrand of $J_r$. Defining $\xi = \frac{1}{2}(a+b)$ and $\eta = \frac{1}{2}(a-b)$ and making a replacement of $r$ as $r = \xi + \eta\cos\phi$, the product $(a-r)(r-b)$ takes a simple form $\eta^2 \sin^2\phi$ making the integral easily tractable. We find

$$\frac{\partial J_r}{\partial r} = -\sqrt{\frac{2m}{|E|}} \int_0^\pi (\xi + \eta\cos\phi) d\phi = -\pi\mu\sqrt{\frac{m}{2}} \frac{1}{|E|^{\frac{3}{2}}}$$

which yields for $J_r$

$$J_r = \lambda + \pi\mu\sqrt{\frac{2m}{|E|}}$$

where $\lambda$ is a constant.

$\lambda$ can be determined by considering a circular orbit of radius $a$ (and $p_r = 0$) when the Hamiltonian is given by

$$H = \frac{1}{2m}(p_r^2 + \frac{\alpha_\theta^2}{r^2}) - \frac{\mu}{r}$$

[1] R. Hosley, Lecture notes on Hamiltonian dynamics (2014), University of Edinburgh, unpublished.

Hamiltonian equation $\dot{p_r} = \frac{\partial H}{\partial r} = 0$ gives the value of $a = \frac{\alpha_\theta^2}{m\mu}$ using the above form of $H$. Since the Hamiltonian gives the energy of the system, substituting this value of $a$ in $H$ implies $E = -\frac{1}{2}\frac{m\mu^2}{\alpha_\theta^2}$. We are now in a position to estimate $\lambda$ by setting $J_r = 0$ since $p_r = 0$. Using the just determined value of $E$ we find $\lambda = -\alpha_\theta$. Hence we obtain $J_r$ as

$$J_r = -2\pi\alpha_\theta + \pi\mu\sqrt{\frac{2m}{|E|}}$$

We refer the readers to the Goldstein book (see bibliography at the end) for a more improved method, involving complex integrations, for the evaluation of $J_r$.

Finally, let us turn to $J_\theta$ which reads

$$J_\theta = \oint p_\theta d\theta = \oint \frac{dW_\theta}{d\theta} = \oint \sqrt{\alpha_\theta^2 - \alpha_\phi^2 \text{cosec}^2\theta} d\theta$$

One way to tackle the integral is to call the ratio $\frac{\alpha_\phi}{\alpha_\theta} = \cos\chi$ and rewrite $J_\theta$ as

$$J_\theta = \alpha_\theta \oint \sqrt{1 - \cos^2\chi \text{cosec}^2\theta} d\theta$$

For the closed path we define the circuit as it is traced out by $\theta$ running from say, $-\theta_0$ to $\theta_0$ and back. The value of $\theta_0$ is furnished by the equation $\sin\theta_0 = \cos\chi$ if we look at the vanishing condition of the integrand corresponding to such a value of $\theta$. In other words, $\theta_0$ is determined from $\chi$ as $\theta_0 = \frac{\pi}{2} - \chi$. In this way, $J_\theta$ is completely given by

$$J_\theta = 4\alpha_\theta \int_0^{\theta_0} \sqrt{1 - \cos^2\chi \text{cosec}^2\theta} d\theta$$

where the factor of 4 occurs because by symmetry it suffices to integrate over the range $(0, \theta_0)$ only.

The rest is straightforward. We make a change of variable $\cos\theta = \sin\chi \sin\psi$ which implies that for $\theta = \frac{\pi}{2}, \psi = 0$ while for $\theta = 0, \psi = \frac{\pi}{2}$. Note that $\theta = 0$ implies $\chi = \frac{\pi}{2}$ so that the integrand in the expression for $J_\theta$ does not become singular.

In this way $J_\theta$ acquires the form

$$J_\theta = 4\alpha_\theta \int_0^{\frac{\pi}{2}} \sqrt{1 - \cos^2\chi \text{cosec}^2\theta} d\theta$$

It is left as an exercise to evaluate the last integral (*Hint*: put $t = \tan\psi$). The final form of $J_\theta$ is

$$J_\theta = 2\pi(\alpha_\theta - \alpha_\phi)$$

It is interesting to observe that when we take the sum $J_r + J_\theta + J_\phi$ all the $\alpha$'s cancel out and we are left with the following energy relation:

$$E = -\frac{2m\pi^2\mu^2}{(J_r + J_\theta + J_\phi)^2}$$

where recall that we are considering bound state or periodic motion for which $E < 0$. The corresponding frequencies $\nu_r$, $\nu_\theta$ and $\nu_\phi$ are the same and given by the partial derivative of the respective $J$ with the corresponding $\nu$:

$$\nu_r = \nu_\theta = \nu_\phi = \frac{4m\pi^2\mu^2}{(J_r + J_\theta + J_\phi)^3}$$

which is completely symmetric in $J_r, J_\theta, J_\phi$.

### 9.7.1 Motion of a particle in a 2-dimensional rectangular well

Let us now investigate[2] the nature of the trajectories when the particle is confined in an infinite rectangular well, with center at $(0,0)$ and vertices at $(\pm a\ ,\ \pm b)$. The two-dimensional potential is given by:

$$V(x,y) = \begin{cases} 0 & |x| < a \quad \text{and} \quad |y| < b \\ \infty & |x| > a \quad \text{and} \quad |y| > b \end{cases} \tag{9.81}$$

Employing a Type-II generating function, with the assumption that the Hamiltonian $H$ does not depend on $t$ explicitly, gives the following equation for the characteristic function $W$:

$$\frac{1}{2m}\left[\left(\frac{\partial W}{\partial x}\right)^2 + \left(\frac{\partial W}{\partial y}\right)^2\right] = E \tag{9.82}$$

Writing $W$ in a separated form, namely, $W = X(x) + Y(y)$, we obtain the following differential equations:

$$\left(\frac{dX}{dx}\right)^2 = \alpha_x^2, \quad \left(\frac{dY}{dy}\right)^2 = \alpha_y^2 \tag{9.83}$$

with $E = \dfrac{1}{2m}\left(\alpha_x^2 + \alpha_y^2\right)$. This implies for the momenta

$$p_x = \pm\alpha_x, \quad p_y = \pm\alpha_y \tag{9.84}$$

---

[2]B. Bagchi, S. Mallik and C. Quesne, Infinite square well and periodic trajectories in classical mechanics, preprint arXiv:physics/0207096 (2002); B. Bagchi and A. Sinha, A classification of classical billiard trajectories, preprint arXiv:physics/09074892 (2009).

In the above the signs show the reversal in the direction of motion of the particle each time it hits the barriers at $x = \pm a$ and $y = \pm b$.

The action variables can now be calculated easily and turn out to be

$$J_x = 4a\alpha_x \tag{9.85}$$

$$J_y = 4b\alpha_y \tag{9.86}$$

so that $E$ takes the form

$$E = \frac{1}{32m}\left(\frac{J_x^2}{a^2} + \frac{J_y^2}{b^2}\right) \tag{9.87}$$

Thus the natural frequencies of the system are obtained to be

$$\nu_x = \frac{\partial E}{\partial J_x} = \frac{J_x}{16ma^2}, \quad \nu_y = \frac{\partial E}{\partial J_y} = \frac{J_y}{16mb^2} \tag{9.88}$$

which, with the help of (9.85) and (9.86) read equivalently

$$\nu_x = \frac{p_x}{4ma}, \quad \nu_y = \frac{p_y}{4mb} \tag{9.89}$$

Thus the natural frequencies are functions of the particle velocity and the dimensions of the well. Consequently, three types of trajectories are possible for the particle trapped in the rectangular well, as discussed below.

---

## 9.8 Possible trajectories

In this section we shall discuss in detail three possible types of trajectories of the trapped particle, namely,
1. Periodic trajectories
2. Open trajectories
3. Special trajectories when the billiard ball hits a corner

### 9.8.1 Periodic trajectories

To study periodic or closed trajectories we note that for the particle to execute periodic motion, it must return to the starting point with its initial momenta after a certain duration of time. This is possible only if

$$T = n_x T_x = n_y T_y \tag{9.90}$$

where $T_x$ and $T_y$ represent the time in which the particle reaches the starting point with its initial momenta in the $x$ and $y$ directions, respectively, $T$ is the time period of the orbit, and $n_x$ , $n_y$ are integers. Thus, if the particle starts from the origin at an angle $\theta$ to the $x$ direction, where

$$\tan\,\theta = \frac{p_y}{p_x} \tag{9.91}$$

then for closed orbits tan $\theta$ must be rational,[3] with the time period given by (9.90). We can straightforwardly rearrange $\tan\theta$ to write down

$$\tan\,\theta = \frac{b\nu_y}{a\nu_x} = \frac{bT_x}{aT_y} = \frac{bn_y}{an_x} \tag{9.92}$$

We now consider the following illustrative examples with the particle starting from the origin in each case.

#### 9.8.1.1 Some explicit examples for periodic trajectories

**Case 1 :** $n_y = 1, n_x = 4$ **, i.e.,** $\dfrac{p_y}{p_x} = \dfrac{b}{4a}$

The trajectories traced out by the particle are given by the following results for the intervals of time specified:

[3]We consider those cases where the linear momentum in the $x$ and $y$ directions, namely, $p_x$ , $p_y$ are expressible in rational form, so that the action variables $J_x$ , $J_y$ as well as the natural frequencies $\nu_x$ , $\nu_y$ are also rational.

$$
\begin{array}{lll}
x = \dfrac{p_x}{m} t & y = \dfrac{p_y}{m} t & t_{16(r-1)} \leq t \leq t_{16r-15} \\
x = -\dfrac{p_x}{m}(t - t_{16r-15}) + a & y = \dfrac{p_y}{m}(t - t_{16r-15}) + \dfrac{b}{4} & t_{16r-15} \leq t \leq t_{16r-14} \\
x = -\dfrac{p_x}{m}(t - t_{16r-14}) & y = \dfrac{p_y}{m}(t - t_{16r-14}) + \dfrac{b}{2} & t_{16r-14} \leq t \leq t_{16r-13} \\
x = \dfrac{p_x}{m}(t - t_{16r-13}) - a & y = \dfrac{p_y}{m}(t - t_{16r-13}) + \dfrac{3b}{4} & t_{16r-13} \leq t \leq t_{16r-12} \\
x = \dfrac{p_x}{m}(t - t_{16r-12}) & y = -\dfrac{p_y}{m}(t - t_{16r-12}) + b & t_{16r-12} \leq t \leq t_{16r-11} \\
x = -\dfrac{p_x}{m}(t - t_{16r-11}) + a & y = -\dfrac{p_y}{m}(t - t_{16r-11}) + \dfrac{3b}{4} & t_{16r-11} \leq t \leq t_{16r-10} \\
x = -\dfrac{p_x}{m}(t - t_{16r-10}) & y = -\dfrac{p_y}{m}(t - t_{16r-10}) + \dfrac{b}{2} & t_{16r-10} \leq t \leq t_{16r-9} \\
x = \dfrac{p_x}{m}(t - t_{16r-9}) - a & y = -\dfrac{p_y}{m}(t - t_{16r-9}) + \dfrac{b}{4} & t_{16r-9} \leq t \leq t_{16r-8} \\
x = \dfrac{p_x}{m}(t - t_{16r-8}) & y = -\dfrac{p_y}{m}(t - t_{16r-8}) & t_{16r-8} \leq t \leq t_{16r-7} \\
x = -\dfrac{p_x}{m}(t - t_{16r-7}) + a & y = -\dfrac{p_y}{m}(t - t_{16r-7}) - \dfrac{b}{4} & t_{16r-7} \leq t \leq t_{16r-6} \\
x = -\dfrac{p_x}{m}(t - t_{16r-6}) & y = -\dfrac{p_y}{m}(t - t_{16r-6}) - \dfrac{b}{2} & t_{16r-6} \leq t \leq t_{16r-5} \\
x = \dfrac{p_x}{m}(t - t_{16r-5}) - a & y = -\dfrac{p_y}{m}(t - t_{16r-5}) - \dfrac{3b}{4} & t_{16r-5} \leq t \leq t_{16r-4} \\
x = \dfrac{p_x}{m}(t - t_{16r-4}) & y = \dfrac{p_y}{m}(t - t_{16r-4}) - b & t_{16r-4} \leq t \leq t_{16r-3} \\
x = -\dfrac{p_x}{m}(t - t_{16r-3}) - a & y = \dfrac{p_y}{m}(t - t_{16r-3}) - \dfrac{b}{4} & t_{16r-3} \leq t \leq t_{16r-2} \\
x = -\dfrac{p_x}{m}(t - t_{16r-2}) & y = \dfrac{p_y}{m}(t - t_{16r-2}) - \dfrac{b}{2} & t_{16r-2} \leq t \leq t_{16r-1} \\
x = \dfrac{p_x}{m}(t - t_{16r-1}) - a & y = \dfrac{p_y}{m}(t - t_{16r-1}) - \dfrac{b}{4} & t_{16r-1} \leq t \leq t_{16r}
\end{array}
\tag{9.93}
$$

where $r = 1, 2, \cdots$ and $t_n = (2n-1)\dfrac{ma}{p_x}$ , $n = 1, 2, \cdots$

Thus, in this case, the trajectories are periodic when the velocities of the particle in the $x$ and $y$ directions are such that the time taken by the particle to cover the distance $a$ in the $x$ direction, is the same as the time it takes to cover the distance $b/4$ in the $y$ direction:

$$t_1 = \frac{ma}{p_x} = \frac{mb}{4p_y}$$

This gives the time period as $T = 4T_x = T_y$, where $T_x = 4t_1$ and $T_y = 16t_1$.

**Case 2 :** $n_y = 2, n_x = 3$ **, i.e.,** $\frac{p_y}{p_x} = \frac{2b}{3a}$

The particle can be shown to trace out the following trajectories:

$$\begin{aligned}
&x = \frac{p_x}{m}t && y = \frac{p_y}{m}t && t_{12(r-1)} \le t \le t_{12r-11}\\
&x = -\frac{p_x}{m}(t - t_{12r-11}) + a && y = \frac{p_y}{m}(t - t_{12r-11}) + \frac{2b}{3} && t_{12r-11} \le t \le t_{12r-10}\\
&x = -\frac{p_x}{m}(t - t_{12r-10}) + \frac{a}{2} && y = -\frac{p_y}{m}(t - t_{12r-10}) + b && t_{12r-10} \le t \le t_{12r-9}\\
&x = \frac{p_x}{m}(t - t_{12r-9}) - a && y = -\frac{p_y}{m}(t - t_{12r-9}) && t_{12r-9} \le t \le t_{12r-8}\\
&x = \frac{p_x}{m}(t - t_{12r-8}) + \frac{a}{2} && y = \frac{p_y}{m}(t - t_{12r-8}) - b && t_{12r-8} \le t \le t_{12r-7}\\
&x = -\frac{p_x}{m}(t - t_{12r-7}) + a && y = \frac{p_y}{m}(t - t_{12r-7}) - \frac{2b}{3} && t_{12r-7} \le t \le t_{12r-6}\\
&x = -\frac{p_x}{m}(t - t_{12r-6}) && y = \frac{p_y}{m}(t - t_{12r-6}) && t_{12r-6} \le t \le t_{12r-5}\\
&x = \frac{p_x}{m}(t - t_{12r-5}) - a && y = \frac{p_y}{m}(t - t_{12r-5}) + \frac{2b}{3} && t_{12r-5} \le t \le t_{12r-4}\\
&x = \frac{p_x}{m}(t - t_{12r-4}) - \frac{a}{2} && y = -\frac{p_y}{m}(t - t_{12r-4}) - b && t_{12r-4} \le t \le t_{12r-3}\\
&x = -\frac{p_x}{m}(t - t_{12r-3}) + a && y = -\frac{p_y}{m}(t - t_{12r-3}) && t_{12r-3} \le t \le t_{12r-2}\\
&x = -\frac{p_x}{m}(t - t_{12r-2}) - \frac{a}{2} && y = \frac{p_y}{m}(t - t_{12r-2}) - b && t_{12r-2} \le t \le t_{12r-1}\\
&x = \frac{p_x}{m}(t - t_{12r-1}) - a && y = \frac{p_y}{m}(t - t_{12r-1}) - \frac{2b}{3} && t_{12r-1} \le t \le t_{12r}
\end{aligned} \tag{9.94}$$

where $r = 1, 2, \cdots$ and $t_n = (2n-1)\dfrac{ma}{p_x}$ , $n = 1, 2, \cdots$

It is easy to observe that in this case $t_2 = 3t_1/2,\ t_3 = 3t_1,\ t_4 = 9t_1/2,\ t_5 = 5t_1$, etc. Thus, $T_x = t_5 - t_1 = 4t_1$ and $T_y = t_8 - t_2 = 6t_1$, where $t_1 = \dfrac{ma}{p_x} = \dfrac{2mb}{3p_y}$, giving $T = 12t_1 = 3T_x = 2T_y$.

### 9.8.2 Open trajectories

If the initial angle $\theta$ is such that $\dfrac{p_y}{p_x}$ is irrational, then the orbit is an open one. This is due to the fact that the time periods in the $x$ and $y$ directions are such that one cannot find integral values of $n_x$, $n_y$ satisfying equation (9.90). We have traced out such a trajectory for $\theta = 30^0$. Even after multiple reflections from the perfectly elastic walls of the rectangular well, the orbit does not close.

### 9.8.3 Special trajectories when the billiard ball hits a corner

We now address the interesting case of special non-periodic trajectories when the particle hits one of the corners of the billiard table and gets pocketed. For the ball to hit either the right or left wall, the distance travelled in the $x$ direction is $(4n \pm 1)a$, where $n$ is an integer. Similarly, to hit the top or bottom wall the distance travelled in the $y$ direction is $(4m \pm 1)b$, where $m$ is also an integer. If the ball hits a corner, then these two conditions must be satisfied simultaneously. Thus, if the particle hits a corner in time $t$, then

$$\frac{v_y t}{v_x t} = \frac{(4m \pm 1)\,b}{(4n \pm 1)\,a} \tag{9.95}$$

and the condition for the billiard ball to get pocketed reduces to

$$\frac{p_y}{p_x} = \frac{(4m \pm 1)\,b}{(4n \pm 1)\,a} \tag{9.96}$$

If the numerator has positive (negative) sign in (9.96), then the ball hits one of the two corners where $y$ is positive (negative). Similarly, if the denominator has positive (negative) sign in (9.96), then the ball hits one of the two corners where $x$ is positive (negative). From equations (9.91), (9.92) and (9.96), it is obvious that this occurs for odd integral values of both $n_x$ and $n_y$. Based on this we summarize in Table 9.1 the corner in which the ball will get pocketed. Note that we assign the following numbers to the respective corners : $(a, b)$ as corner 1, $(-a, b)$ as corner 2, $(-a, -b)$ as corner 3, and $(a, -b)$ as corner 4. It is observed that the predicted corners are in fact the actual ones.

**Table 9.1 Corners Where the Ball is Pocketed.**

| $n_y/n_x$ | Sign (numerator, denominator) | Corner |
|---|---|---|
| 1/3 | $(-,+)$ | 2 |
| 1/5 | $(+,+)$ | 1 |
| 1/7 | $(-,+)$ | 2 |
| 1/9 | $(+,+)$ | 1 |
| 3/7 | $(-,-)$ | 3 |
| 5/7 | $(-,+)$ | 2 |
| 3/5 | $(+,-)$ | 4 |
| 5/9 | $(+,+)$ | 1 |
| 7/9 | $(+,-)$ | 4 |

## 9.9 Summary

Since the choices of generalized coordinates in most cases are rather arbitrary it is sometimes necessary to look for transformations that ease the mathematical complications confronting a given physical problem. Canonical transformations form one class of transformations that offer a simplified way of looking at the physical properties of a mechanical system through equations that can be represented in simple terms. Looking for invariance of Hamilton's equations under a set of suitable transformations does indeed give the right clue in this direction. For one thing we can equivalently think of those class of transformations that preserve the invariance of Poisson brackets defined in terms of new variables. For another, the symplectic character of canonical transformations holds equivalently. Permuting the set of generalized coordinates and the set of generalized or conjugate momenta one can define four different types of functions that generate these transformations. Such functions are called generating functions. The extreme possibility of the transformed variables becoming explicitly absent in the transformed Hamiltonian gives us the Hamilton–Jacobi equation. This has many rich consequences — the most noteworthy being the formulation of action-angle variables. This chapter discussed through examples the underpinnings of canonical transformations and the new directions they open up.

# Exercises

1. Show that the following transformations of coordinates are canonical:

$$(i)\quad Q = \sqrt{q}\cos 2p, \quad P = \sqrt{q}\sin 2p$$

$$(ii)\quad Q = \tan^{-1}(\frac{q}{p}), \quad P = \frac{1}{2}(q^2 + p^2)$$

$$(iii)\quad Q = \log(\frac{\sin p}{q}), \quad P = q\text{cot}p$$

$$(iv)\quad Q = pq^2, \quad P = q^{-1}$$

$$(v)\quad Q = \log(1 + \sqrt{q}\cos p), \quad P = 2(1 + \sqrt{q}\cos p)\sqrt{q}\sin p$$

2. What is the canonical transformation generated by $G_1 = \lambda q^2 \text{cot}Q$, where $\lambda$ is a constant?

3. Deduce a canonical transformation such that the Hamiltonian $H$ of the form $H = p^2 + (4q+1)p + 4q^2 - 2q$ is coverted to $K(Q,P) = \frac{1}{2m}P^2 - mgQ$ where $m$ and $g$ are constant quantities. Interpret $K(q,P)$.

4. Show that the set of transformations

$$Q = \sqrt{\frac{2q}{k}}\cos p, \quad P = \sqrt{2qk}\sin p$$

is canonical and the generating function $G_1(q,Q)$ is

$$G_1(q,Q) = \frac{1}{2}Q\sqrt{2qk - k^2Q^2} - q\cos^{-1}(Q\sqrt{\frac{k}{2q}})$$

5. Prove that the generating function facilitating transformation from Cartesian to spherical coordinates through the usual relations

$$x = r\sin\theta\cos\phi, \quad y = r\sin\theta\sin\phi, \quad z = r\cos\theta$$

is given by the Type II form

$$G_2(p,Q) = -(r\sin\theta\cos\phi p_x + r\sin\theta\sin\phi p_y + r\cos\theta p_z)$$

6. Let a Hamiltonian be given by

$$H = \frac{1}{2}(p_x^2 + p_y^2) + c^2(q_x^2 + q_y^2)$$

If we transform to new variables defined by

$$Q_x = q_x^2 + \frac{p_x^2}{c^2}, \quad Q_y = \frac{1}{2}c^2[(q_x^2 + q_y^2) + \frac{1}{c^2}(p_x^2 + p_y^2)]$$

prove that the transformation is canonical and the transformed Hamiltonian is given by simply $Q_y$.

7. Show that the canonical transformation mapping the variables $(q_x, p_x, q_y, p_y)$ to $(Q_x, P_x, Q_y, P_y)$ where the latter are given by

$$Q_x = q_x cos\theta - q_y \sin\theta, \quad Q_y = q_x \sin\theta + q_y \cos\theta, \quad P_x = p_x cos\theta - p_y \sin\theta,$$
$$P_y = p_x \sin\theta + p_y \cos\theta$$

can be effected by the generating function $G_2 = (P_x q_x + P_y q_y)\cos\theta + (P_y q_x - P_x q_y)\sin\theta$.
Using such a transformation, show that a coupled oscillator Hamiltonian

$$H = \frac{p_x^2 + \omega_x^2 q_x^2}{2} + \frac{p_y^2 + \omega_y^2 q_y^2}{2} + g q_x q_y$$

gets decoupled in a manner $H \to K$:

$$K = \frac{P_x^2 + \Omega_x^2 Q_x^2}{2} + \frac{P_y^2 + \Omega_y^2 Q_y^2}{2}$$

where

$$\Omega_x^2 = \frac{1}{2}[\omega_x^2 + \omega_y^2 + \sqrt{(\omega_x^2 - \omega_y^2)^2 + 4g^2}], \quad \Omega_y^2 = \frac{1}{2}[\omega_x^2 + \omega_y^2 - \sqrt{(\omega_x^2 - \omega_y^2)^2 + 4g^2}]$$

and $\tan(2\theta) = -\frac{2g}{\omega_x^2 - \omega_y^2}$.

8. The motion of a particle of mass $m$ moving in a plane under a central force is controlled by the Hamiltonian

$$H = \frac{1}{2m}(\dot{r}^2 + r^2\dot{\theta}^2) + V(r)$$

Determine Hamilton's characteristic function and solve the problem completely.

9. If a time-independent generating function $G_2(q, P)$ is given by $G_2(q, P) = qP + \epsilon a q^3 P + \epsilon b q P^3$ where $\epsilon$ is a small quantity and $a$,$b$ are constants, obtain $q$ and $p$ in the forms

$$q = Q - 3\epsilon b Q P^2 - \epsilon a Q^3, \quad p = P + \epsilon b P^3 + 3\epsilon a Q^2 P$$

keeping only up to $O(\epsilon)$ terms. Use the above form for $G_2$, map the anharmonic oscillator Hamiltonian $H = \frac{1}{2}(p^2 + \omega^2 q^2) + \epsilon\beta q^4$ to $K(Q, P) = \frac{1}{2}(P^2 + \omega^2 Q^2) + \epsilon c(P^2 + \omega^2 Q^2)^2$, where $c$ is to be determined. Show that $Q$ and $P$ are in the forms

$$Q = A\cos(\Omega t + \delta), \quad P = A\sin(\Omega t + \delta)$$

where $A$ is a constant and $\Omega = \omega + \frac{3\epsilon\beta A^2}{2\omega}$.

10. For a power law potential $V(q) = \lambda q^{2n}$, where $\lambda > 0$ and $n$ is a positive integer, appearing in the Hamiltonian $H = \frac{p^2}{2m} + V(q)$, find the action variable $I$ in the form

$$E = \lambda^{\frac{1}{n+1}} (\frac{n\pi I}{J_n})^{\frac{2n}{n+1}} (\frac{1}{2m})^{\frac{n}{n+1}}$$

where $E$ is the energy and $J_n$ is given by the integral

$$J_n = \int_0^1 (1-t)^{\frac{1}{2}} t^{\frac{1-2n}{2n}} \, dt$$

# References


1. Arnold, V.J., (1978) *Mathematical Methods of Classical Mechanics*, Springer-Verlag, New York.

2. Arrowsmith, D.K. and Place, C.M., (1992) *Dynamical Systems, Differential Equations, Maps and Chaotic Behavior*, Chapman & Hall/CRC, New York.

3. Berry, M.V., (1989) *Principles of Cosmology and Gravitation*, Cambridge University Press, London.

4. Calkin, M.G.,(1996) *Lagrangian and Hamiltonian Mechanics*, World Scientific Publishing Co. Pte. Ltd.

5. Corinaldesi, E., (1998) *Classical Mechanics For Physics Graduate Students*, World Scientific Publishing Co. Pte. Ltd.

6. Dirac, P.A.M., (1964) Lectures on Quantum Mechanics, Belfer Graduate School of Science, Yeshiva University.

7. Elsgatts, L., (1975) *Differential Equations and the Calculus of Variations*, Mir Publishers, Moscow.

8. Fasano, A. and Marmi, S., (2002) *Analytical Mechanics*, Oxford University Press, Oxford.

9. Fetter, A.L., and Walecka, J.D., (1980) *Theoretical Mechanics of Particles and Continua*, McGraw-Hill Companies, Inc., New York.

10. Fring, A., (2005) *Lecture Notes on Dynamical Systems*, City University, London.

11. Gantmacher, F., (1970) *Lectures in Analytical Mechanics*, Mir Publishers, Moscow (1975).

12. Goldstein, H., Poole, C. and Safko, J., (2000) *Classical Mechanics*, Addison Wesley, San Francisco, CA.

13. Greiner, W., (2010) *Classical Mechanics: Systems of Particles and Hamiltonian Dynamics*, Springer-Verlag, New York.

14. Gupta, S.N.,(1970) *Classical Mechanics*, Meenakshi Prakashan, Meerut.

15. Hand, L.N. and Finch, J.D., (1998) *Analytical Mechanics*, Cambridge University Press, London.

16. Hilborn, R.C., (1994) *Chaos and Nonlinear Dynamics: An Introduction for Scientists and Engineers*, Oxford University Press, Oxford.

17. Iro, H., (2002) *A Modern Approach to Classical Mechanics*, World Scientific Publishing Co. Pte. Ltd.

18. Jones, D.S., Plank, M. and Sleeman, B.D., (2009) *Differential Equations and Mathematical Biology*, CRC/Taylor & Francis Group, Boca Raton, FL.

19. Kleppner D., and Kolenkov, R., (1973) *An Introduction to Mechanics*, McGraw-Hill Companies, Inc., New York.

20. Landau, L.D., and Lifshitz, E.M., (1976) *Mechanics*, Pergamon Press, Oxford.

21. Lass, H.,(1950) *Vector and Tensor Analysis*, McGraw-Hill Book Company, Inc., New York.

22. Loney, S.L., (1982) *An Elementary on the Dynamics of a Particle and of Rigid Bodies*, Macmillan, Calcutta.

23. Pierre, K., (1994) *Lectures Notes on the Introduction to the Theory of Dynamical Systems*, University of Brussels.

24. Rana, N.C., and Joag, P.S. (1991) *Classical Mechanics*, McGraw Hill Education (India) Private Limited, New Delhi.

25. Raychaudhuri, A.K., *Classical Mechanics—A Course of Lectures*, Oxford University Press, Calcutta.

26. Sagdeev, R.Z., Usikov, D.A., and Zaslavsky, G.M., (1988) *Nonlinear Physics: From the Pendulum to Turbulence to Chaos*, Harwood Academic Publishers, Pennsylvania.

27. Scheck, F., (1994) *Mechanics: From Newton's Laws to Deterministic Chaos*, Speinger-Verlag, Heidelberg.

28. Simmons, G.F., (1974) *Differential Equations*, Tata-McGraw-Hill, New York.

29. Sommerfeld, A., (1952) *Lextures on Theoretical Physics-Mechanics*, Academic Press, Inc., New York.

30. Spigel, M.R., (2006) *Theoretical Mechanics*, Tata-McGraw-Hill Education Private Limited, New Delhi

31. Strogatz, S.H., (2007) *Nonlinear Dynamics and Chaos*, Levant Books, Kolkata.

32. Symon, K.R., (1971) *Mechanics*, Addison-Wesley Publishing Company, Inc., Philippines.

33. Whittaker, E.T. (1988) *A Treatise on the Analytical Dynamics of Particles and Rigid Bodies*, Cambridge University Press, London.

# Index